Sock it to 'em Baby

Forward Air Controller in Vietnam

Garry Gordon Cooper DFC
Robert Ross Hillier BSc

ALLEN&UNWIN

First published in 2006

Copyright © Garry Cooper and Robert Hillier 2006

All rights reserved. No part of this book may be reproduced or transmitted in any form or by any means, electronic or mechanical, including photocopying, recording or by any information storage and retrieval system, without prior permission in writing from the publisher. The *Australian Copyright Act 1968* (the Act) allows a maximum of one chapter or 10 per cent of this book, whichever is the greater, to be photocopied by any educational institution for its educational purposes provided that the educational institution (or body that administers it) has given a remuneration notice to Copyright Agency Limited (CAL) under the Act.

Allen & Unwin
83 Alexander Street
Crows Nest NSW 2065
Australia
Phone: (61 2) 8425 0100
Fax: (61 2) 9906 2218
Email: info@allenandunwin.com
Web: www.allenandunwin.com

National Library of Australia
Cataloguing-in-Publication entry:

Cooper, Gary, 1938- .
  Sock it to 'em baby: forward air controller in Vietnam.

  Bibliography.
  Includes index.
  ISBN 978 1 74114 849 7

  1. Cooper, Gary, 1938- . 2. Vietnamese Conflict, 1961-1975
  - Biography. 3. Soldiers - Australia - Biography. 4. Air
  pilots - Australia - Biography. I. Hillier, Robert, 1938- .
  II. Title.

629.13092

Set in 11.5/13 pt Bembo by Midland Typesetters, Australia
Printed and bound in Australia by Griffin Press

10 9 8 7 6 5 4 3 2

FOREWORD

by Lieutenant General Julian J. Ewell, (ret.)

As the Commanding General of the 9th Infantry Division in the Delta area of Vietnam in 1968 and 1969, and later, as Commander of II Field Force, Vietnam, I was involved in combat operations at the height of the Vietnam War. The 9th Division was generally responsible for combat operations and pacification support from Saigon south to the Mekong River.

In examining the Vietnamese War, particularly in its early stages, there were many situations in which the Communists, although not successful overall, were able to give the Allies a hard time. The set-piece ambush, the set-piece attack on an isolated post, the seizure and defence of a populated area, and the development of a fortified village were all examples of operations which the Communists executed many times and were difficult to handle without incurring excessive friendly military casualties and, in some cases, undue civilian casualties or damage. The Communist style of making war was inherently destructive to the people and physical resources of a country. The Communists 'liberated' the countryside by destroying roads and bridges. They controlled the country by breaking up its social and governmental structure and applying force and terror against the people. By seizing inhabited areas, by fortifying villages, they forced the government into heavy combat which harmed the people and destroyed civilian resources. Their organization of the masses in support of their activities absorbed considerable manpower and economic resources, thereby slowing economic and political progress.

An aggressive and skilful Allied effort captured the initiative and countered the enemy activities with low casualties and little harm to the civilian population. A casual observer of the Vietnamese War would be so conditioned by reading dramatized newspaper accounts of the war that he would visualize Vietnam disappearing under smoke and flame of bombs and artillery shells. However, if one looked at the facts, it was quite apparent that the more the Allied side gained control of the war, the less destructive it became. The Allied effort served to protect the people from Communist inroads and to rebuild the country.

Our air support was furnished by the U.S. 7th Air Force and usually controlled by airborne Forward Air Controllers. One of the Forward Air Controllers assigned to us for much of the period during which I was Commander of the 9th Infantry Division was Flight Lieutenant Garry Cooper of the Royal Australian Air Force. As an aside, the major Australian contribution to the overall war effort was an Infantry Brigade called 'The Australian Task Force'.

Garry Cooper became a hero in the eyes of the entire 9th Division for his gallant actions on the 18th August 1968. On that date, the pilot of the Command Helicopter was killed outright and Flight Lieutenant Cooper seized the controls from the dead pilot preventing an inevitable fatal contact with the ground. The Brigade Commander, a passenger in the helicopter, was hit in the

FOREWORD

neck with a bullet which also struck Flight Lieutenant Cooper's flying helmet. The crash occurred in the late afternoon between friendly and enemy forces some 200 metres in front of the enemy position. Under fire, Flight Lieutenant Cooper assisted the Brigade Commander to relative safety, repelled enemy troops, and killed at least 10 of them at close range. Flight Lieutenant Cooper exhausted his ammunition while covering the Brigade Commander as he was being hoisted aboard another helicopter. With a now empty but useful weapon, Flight Lieutenant Cooper defended himself against attack by two enemy soldiers whom he killed thereby saving the wounded Brigade Commander.

Immediately after Flight Lieutenant Cooper's extraordinary action at great risk to his own life, I recommended him for the award of the *Medal of Honor*, the United States' highest award for valor in combat operations. Unfortunately, the recommendation ran into a host of regulatory problems but to this day, I am still trying to see what can be done about it.

Through the years, I have seen Garry and his wife Jean on their frequent visits to the United States during which we talk together about our combat days in Vietnam. It is an honor to salute Garry Cooper, an Australian who has strengthened the bond between our two countries.

Lieutenant General Julian J. Ewell, (ret)
Fort Belvoir, Virginia.
August 4, 2003.

FOREWORD

by Air Vice-Marshal J.H. Flemming, (ret.)

THIS STORY IS BOTH inspirational and disturbing. It inspires by recounting the deeds of an airman, valiant in many fields of aviation, but disturbs by the determination of successive bureaucracies to deny him formal recognition of his achievements.

I first met Garry Cooper in 1965 when we were together on a Mirage Conversion course at Williamtown. His previous flying experience was apparent. In flying ability he was outstanding and well ahead of his contemporaries in most aspects of the course. Later as his Commander in No 75 Squadron I was impressed by his attitude towards the task, his loyalty and dedication.

While leading him on an exercise with large drop tanks, he experienced a complete engine failure at low altitude and heavy weight . . . By a brilliant display of airmanship, ability and quick thinking, he landed his Mirage on an unused war-time airstrip at Hexham, without damage. It was then that I first learnt of a hierarchical bias which seems to have affected this officer throughout his career.

Vicious rumours started that the forced landing was intentional. It was not until an inspection of the aircraft found that a large eagle hawk had been ingested into the engine, causing a flame out

FOREWORD

and total power loss that the rumours ceased ... I was so impressed by his performance, in saving what should have been the loss of a very valuable aircraft, that I recommended that he be awarded an Air Force Cross.

Not only was he denied the award but I was berated for making the recommendation. The fact that a qualified test pilot, under less hazardous conditions, was awarded the Air Force Cross for landing an unladen Mirage without power, on an operational aerodrome runway, makes such actions by higher authority questionable.

During the next few years I followed Garry Cooper's performance. During this time he saved three more valuable RAAF aircraft from imminent destruction at the risk of his own life. His display of airmanship and flying ability in each instance were worthy of the highest praise but again the hierarchy saw fit to find fault. No note of gratitude or appreciation was ever given. Instead he was castigated and veiled innuendos and rumours abounded.

I have no personal knowledge of his operations as a FAC with the Americans in Vietnam but the enormity of the task and his self effacing descriptions of his day to day life in primitive conditions and operational flying read like a Tom Clancy novel.

Some of the American Tactical Air pilots I have met have only the highest praise for 'Tamale 35'. In the USA he is considered a legend and has been welcomed into their unit associations.

It is easy to understand Cooper's frustration with the bureaucracy in regard to American bravery awards. I fought for over forty years to get the Korean Presidential Citation, awarded to No 77 Squadron in Korea in 1950, accepted by the Australian Government, although not permitted to be worn by those entitled. It was only when, with the help of the Korean Ambassador, we convinced the RAAF that serious embarrassment would incur with an impending visit by the President of South Korea that common sense finally prevailed. The decoration is now worn by all those so entitled.

Despite the many recommendations for bravery awards for Cooper made by the American administration, there is a marked reluctance by the RAAF and the Australian Government to acknowledge or accept any of them. Later the RAAF awarded him the Imperial Distinguished Flying Cross for 'Services in Vietnam'. As his entire operational tour was with the Americans, his actions and operational performance, giving rise to this award, had to be based on American reports. Why there was such opposition by the hierarchy is difficult to understand. Again it would appear that the stigma surrounding his earlier achievements, quite undeserved, carried on to Vietnam.

Garry Cooper has come along way from gliders to captaincy of a Boeing 747. He is a pilot of outstanding ability, determined in peace and valiant in war. During his career he saved at least four valuable aircraft from destruction and earned the gratitude, albeit reluctantly, of the Australian Government and the admiration of our US allies for his service in Vietnam. He should have received accolades not criticism. He deserves better.

I consider it a privilege to be able to read and comment on this book and can recommend it as a valuable contribution to our military history.

Air Vice-Marshal J.H. Flemming
AO, DFC, AM (US), FAIM, FCIT, RAAF Ret.
Canberra
January 2006

CONTENTS

Foreword by Lieutenant General Ewell (ret.)	iii
Foreword by AVM Jim Flemming (ret.)	vi
Acknowledgements	xi
FACs—In Memoriam	xv
Introduction	xvii
1 Flashback	1
2 The early years	22
3 The RAAF and Vietnam	48
4 The May Offensive	70
5 The 'Y' Bridge in Saigon	87
6 The May Offensive continued	114
7 Dong Tam	139
8 Engine failure	165
9 Down in the Boonies	193
10 R and R	220
11 The struggle to get home	246
Appendix—The Medal of Honor issue	257
Glossary	304
Bibliography	312
Index	313

*To my wife Jean, and sons,
Carl, Mark and Ashley, who
make me proud.*

*This book is dedicated to the 220 Forward Air
Controllers and the 2588 soldiers of the
9th Infantry Division, US Army,
who paid the ultimate sacrifice and
did not return to their loved ones.*

ACKNOWLEDGEMENTS

This book has been written for two main reasons. The first is that, although there have been a number of books written from the Australian Army's perspective in Vietnam, few have been written about the RAAF and not one about the Australian forward air controllers (FACs), giving a detailed description of their operations. Chris Coulthard-Clark wrote an excellent book about Australian FACs but it was, in essence, a collection of short personal stories with some history that I will not cover here. My intention in this book is to give an insight into the day-to-day activities of the FACs, particularly operating under the complete control of a foreign power. Throughout the time the Australian FACs served with the United States Air Force (USAF), each individual pilot had a different experience. Even those who were in Vietnam at the same time experienced varied circumstances. It depended a lot on location, enemy activities and just when the FAC was in Vietnam. About a third of my flying was done at night, which I later learned not many of the other Australian FACs experienced. I put in air strikes with 'troops in contact' on 68 occasions, which not many can lay claim to. However, this does not detract from the gallantry all 36 of them displayed during their tours. It takes only one bullet to spoil your day and just flying around Vietnam at low level on a daily basis took a lot of courage.

The second reason is to put the record straight. Over the years I have received considerable ridicule from a number of Australian sources and people who should know better and heard statements based on bar-talk and second-hand stories. Unknown to me, I was recommended for America's highest bravery award, the Medal of Honor, for action occurring on the 18 and 19 August 1968. I found this out when General Ewell, who made the recommendation, asked the Australian Defence Department in 1975 what had happened to his recommendation. He was told the United States could not award me the Medal of Honor because I was a foreign national but to this day his question has not been fully answered honestly. There are seven United States military documents missing from my file that corroborate the actions on the 18 and 19 August 1968. What happened to them? The government's Review, with its predetermined outcome (mentioned in chapter 11 and the appendix), has done nothing to help the innuendo, intrigue and controversy that surround this issue. By trying to cover up their immoral conduct by suggesting the action on 18 August 1968 never happened, despite the statements of many senior US officers, they have provided sceptics with ammunition to further deny and denigrate my service. These sceptics have acted on rumours, without the full facts available to them, and in some cases out of pure jealousy. For some people, to be trained for combat and to not see any develops resentment against those who have. The implementation of the Australian awards system during the Vietnam conflict fostered resentment among our own people. On two occasions I have been forced to take legal action to quell a line of libel levelled at me. I have heard of some people stating that the FACs wrote their own US Awards, which can only be the thoughts of an ignorant person. Any US recommendation had to be accompanied with two or three eye-witness statements. The recipient's immediate superior had to initiate the recommendation based on these statements. The exception to this was with the US Army where a Brigade Commander could make an

ACKNOWLEDGEMENTS

immediate award 'in the field', usually a Bronze Star. The recommendation was then evaluated by the Unit's Awards Branch and verified before being signed by the commanding officer, frequently of full Colonel rank. It was then forwarded to Headquarters in Saigon where it was evaluated and verified again before being approved, or disapproved. Final approval was then given by the Commander of the 7th Air Force. With this process, how could a junior rank, foreign officer write his own awards?

There are many people whom I must thank for their assistance and insistence that this book be published. Firstly I owe a debt of gratitude to Lieutenant General Julian J. Ewell, USA (ret.), who has stood by his award recommendation since 1968. A man of his stature should not have to chase after an award for a foreign, junior officer. But he has, and that shows how dedicated and compassionate he is towards 'his' soldiers. Colonels William L. Walker and Richard F. Nelson, USAF ret., who were my immediate commanders, along with Colonel James T. Patrick all tried to bring my service to the notice of the Australian Government. They are men I admire for much more than this.

To all the members of the 9th Infantry Division and, in particular, the members of 5th Battalion, 60th Infantry, who have accepted me among their gallant ranks as a lifetime friend, I say, it's an honour to be one of you.

I owe a debt of thanks to Robert Hillier, who battered me into writing this book, led me through the procedure and co-wrote many sections.

There have been many people, too many to mention here unfortunately, who have supported me over the years and were an enormous help. Colin Benson, himself a Vietnam veteran, spent hours at research and on the computer trying to bring the facts of my service to the notice of our government. Only recently, due to personal disasters, he has had to reduce his very thankful input but he remains a trusted friend. One in particular who has hung in there is Laurie Schneider, who has worked untiringly on my behalf. Others I want to mention are Fred Kirkland, Bill Connell, Vern Lewis, Vic Cannon, Jim Flemming, Gene Rossel, Don

Handsley and the many others who have given assistance and shown their support.

I would also like to to thank Allen & Unwin for their assistance and guidance in converting my writings into a readable book.

In writing this book, I have relied on my flying logbook, memory and diary. Unfortunately, my diary has been retyped by a number of well-meaning individuals and errors have occurred but through research, subsequent discoveries have been numerous. The memory is fallible and it is possible that some of the dates and events may not have happened in the exact sequence described. And some of the names have been changed for the usual reasons, or made up when I could find no record of them. Nevertheless, I have tried to faithfully draw a picture of my experiences and every effort has been made to accurately describe the incidents in which I was involved.

FACS—IN MEMORIAM

*They were young not yet wise but clear in what was right
and what was fair.
They knew that they in their little grey planes could cover
their troops with a mantle of care.
Their presence alone enough to deter even the worst of
what was there.
They did their duty, their duty of care, eyes in the sky,
ears in the air.
Neither painted for war, nor painted for show,
uncamouflaged they were.
Ringmasters of the greatest show on earth.
Small, grey, guardians alone, and unarmed.
With unlimited power, aerial shields, aerial swords, accurate,
immediate, fatal.
How do they count those analysts of war, account, amount and
total the score, compile and record their statistical war?
BDA, KIA, MIA, KBA.
What does it mean? What does it matter? What matters above all,
Is the unknown number of unknown soldiers who didn't fall,
and who, decades later, still enjoy the love of their families, the
pleasure of friends, the smell of spring, and the song of the
wind, purely because those little grey planes were there
whenever, and wherever, they were needed.
God bless them.*

Flying Officer David P. Robson
FAC 'Jade 07'—Vung Tau—1969

INTRODUCTION

THIS BOOK IS ABOUT A young Australian pilot—Flight Lieutenant Garry Gordon Cooper. He joined the Royal Australian Air Force and became a fighter pilot. In April 1968 he was sent to South Vietnam to fly with the United States Air Force as a forward air controller. He was flown to South Vietnam in a first-class seat on a Qantas Boeing 707. He returned to Australia on 29 October on board a noisy Hercules transport, sitting in the cargo bay strapped to the airframe accompanied by several returning dead in coffins. What a difference a war makes! By the end of 1968 the war was becoming increasingly unpopular in Australia as elsewhere.

When he arrived in Vietnam the communists were starting a counterstrike following the Tet Offensive. Robert McNamara had told the American people that the situation in Vietnam was under control and that it would not be long before the communists were forced to negotiate. McNamara's assurances rang hollow when the communists launched an offensive to coincide with the Vietnamese lunar New Year called *Tet Nguyen Dan*, meaning 'Fete of the First Day'—it is a national family celebration, held sacred and most significant for the Vietnamese people, wherever they may be.

As in all things in life, timing is the essence of success. The Tet Offensive was launched on 30 January 1968 in the north. Eighty thousand communists attacked cities and military targets across South Vietnam. The communists captured the beautiful old city of Hue, situated 75 kilometres south of the demilitarised zone. It took a week for the Americans to recapture it. During that week 2800 Vietnamese with American connections who lived in Hue were executed. The American Embassy in Saigon was attacked and fighting in that city lasted a week.

The South Vietnamese, the American military and its allies, which included Australia's 5 and 6 battalions and Royal Australian Regiments, soundly beat the communists. The North Vietnamese Army, under the command of General Vo Nguyen Giap, and the Viet Cong suffered heavy casualties—General Giap later admitted that they were a spent force. They could not match the firepower of the Americans. Also the uprising of the people of South Vietnam did not occur as communist cadres had expected. It seems contradictory that the Vietnamese, who are gentle and spiritual people, should subscribe to communism. The Vietnamese communists were indoctrinated in Moscow at the height of the Cold War and the people were caught between the hammer of Stalinist communism and the western capitalist anvil.

Despite the Allied victory, the Tet Offensive shook the American psyche and convinced many that the war could not be won. US President Lyndon Johnson became seriously ill and, unfortunately, could not stand for re-election. This was seen wrongly as an admission that his war policies were flawed. The resolve of the men and women in the street wavered. America faltered. The communists seized on America's uncertainty and won a propaganda coup. Someone I read said, 'Giap had snatched victory from the jaws of defeat'. General Giap learned not to take on the Americans in a shootout. By April 1968 the communists had recovered and began launching guerilla sorties against South Vietnam. As Flight Lieutenant Cooper landed in Tan Son Nhut Airport the violence was escalating across the South once again. The young Australian fighter pilot was about to have an adventure that would change his life.

1

FLASHBACK

THE WET RUNWAY GLISTENS in the marker lights as it slips under the lumbering Boeing 747. The acceleration presses the occupants into the back of their seats as the big airliner gathers speed. After a few hundred metres it lifts its 410 passengers and crew clear of Hong Kong's busy Kai Tak Airport, and out across Kowloon Bay through the Lyemun Gap. The landing gear comes up with a thud and the aircraft is heading towards Sydney, Australia, eight hours away. I am the pilot in command and my first officer is the attractive and capable Penny Hanrahan. The equally capable, but not so attractive, flight engineer is Ron Jones. Sitting in the jump seat is my son Mark, who is accompanying me on this, my last international flight. I have been told that I will be returning to fly the Australian domestic routes on a Boeing 767 aircraft. I am rather depressed about this, as I have always enjoyed international flying. However, the company has a questionable policy of senior pilots returning to domestic routes when they reach the age of 60, which I have reached in 1998.

We are now about two hours into the flight and approaching Zamboanga, known as the city of flowers, in the southern Philippines. Ahead we can see towering cumulo-nimbus clouds of tropical thunderstorms that are lit by tremendous bolts of lightning. It is an awesome sight. The clouds rise up some 50 000 feet. The thunderheads tower 17 000 feet higher than our altitude and seem like giant mountains above us. They are too high to climb over so I gently change the course of the 360 tonnes of aeroplane to take us between the storms. The heavy aircraft responds to my touch like some obedient giant flying animal. The cockpit has taken on a surreal ambience as lightning flashes around it as we drone on through. The kaleidoscope of brilliant and changing light patterns cause my thoughts to drift to a time when I experienced similar flashes in a country not too far west of here. These flashes, however, were not an act of nature but the work of men at war.

I remember that I was asleep in a prefabricated wooden hut in Vietnam. There were thousands of these constructions on military facilities in the country. It is July 1968 and I shared this hut with four others. We were resting between combat missions and each of us had a small piece of territory in the hut marked off by steel lockers and empty ammo cases. We stored our meager belongings in these cold steel receptacles. The hut was situated inside the wire of the United States Army Base at Dong Tam in the Republic of South Vietnam. Being inside the wire meant the perimeter of the base was secured by the soldiers whose home it was. At present Dong Tam was base camp to the United States Army's 9th Infantry Division. I was the only Australian in this hut. My four companions were United States Air Force (USAF) fighter pilots. I was also a fighter pilot but served with the Royal Australian Air Force. My parent unit was the USAF 19th Tactical Air Support Squadron, usually referred to as 19TASS and I have been seconded to the USAF as a forward air controller. Forward air controllers were known in Vietnam by the acronym FACs. The USAF had assigned me to the 3rd Brigade, 9th Infantry Division, of the US Army.

That night I was sleeping rather soundly, aided by the sedative effect of fatigue and the half-a-dozen Bacardi rum and cokes I drank before dropping into bed. I had tied my washed flight suit to the ceiling fan so it would dry quickly in the rhythmic circular motion of the blades, and also so I could find it easily in the dark. Sometimes during the night, the noise of incoming mortar and rocket fire would wake us and send us scurrying to the safety of a nearby bunker. However, that night the duty sergeant radio operator of my tactical air control party (TACP) quietly entered the hut so as not to wake my companions, came to my bunk and whispered urgently, 'Sir . . . Captain Cooper, we have troops-in-contact. Sir, let's go!'

Troops-in-contact are friendly troops engaged in combat with enemy forces. I dragged myself back to the conscious world of this unwanted reality. In the quiet semi-darkness I muttered something incomprehensible in acknowledgement of his urgent request. Twenty minutes later I was rattling along the Dong Tam perforated steel plate (PSP) runway in a Cessna 0-1E monoplane, slowly gathering speed. In the back seat of the aircraft was a newly arrived American fighter pilot whom I was training in combat operations—Major Richard 'Dick' Nelson from Arkansas. He was assigned as the air liaison officer (ALO) of the 3rd Brigade, taking over from Major William 'Bill' Walker. My mind was focused on the mission ahead. I was driven to get to the location quickly because soldiers might have already been dying. The objective of my mission is simple. It is to coordinate air strikes to help units of the 5th Battalion, 60th Infantry of the 9th Infantry Division, who were engaged in a vicious firefight with enemy Viet Cong (VC) forces. The implementation of air strikes is complex. My role as a forward air controller in the impending operation was to direct the placing of ordnance on the enemy without endangering the lives of friendly forces.

I had the engine of the little aircraft at full throttle as it climbed away from the airfield into the black sky. All combatants in Vietnam know the little aircraft I was flying as a 'Birddog'. It seems ridiculous to have such a puny aircraft involved in a high-tech war. It carries next to no armament, its speed is only about 100 knots (185 kph), and it can be shot easily out of the sky. Apart from its USAF markings, it is an all-over light grey colour and has eight 2.75-inch (70 mm) smoke marker rockets slung in two pods of four under each wing. The rockets are known as 'Willy Petes'. To the casual observer I could be a weekend fly-in but appearances are deceptive. The Birddogs regularly wreck havoc on the enemy. The VC and the North Vietnamese Army (NVA) fear the little aircraft, knowing that ordnance can be dropped on them if the FAC directs it. FAC pilots display few of the affectations usually associated with combat fighter pilots. There are no polka-dot scarves, no cobras painted on their aircraft like the Air Cavalry chopper pilots sometimes have. All forward air controllers wear a sidearm that is usually a military issue .38 six-shot Colt revolver worn in a low-slung holster to keep it out of the way while flying.

As we cleared the end of the runway the night silence was interrupted by a burst of machine-gun fire aimed at the little aircraft from the jungle below. I could see the green tracers from the cockpit arching up but falling away behind us as we climbed into the darkness. At night everyone outside the wire was the VC but my rookie in the back seat and I had bigger fish to fry, so the brazen tracer burst from the jungle below was disdainfully ignored. Dick Nelson was a veteran F86 Sabre fighter pilot from the Korean War and my task as the unit's combat instructor pilot (CIP) was to teach him the ropes. By now he may have been wondering if it was such a good idea to become a forward air controller.

I was by now fully awake and planning my night's work. I would have enjoyed a cup of coffee before getting into the air but lives were at risk and seconds can mean the difference between life and death for my soldiers. My call sign was the ridiculous 'Tamale 35'—*Tamale* is a hot Mexican dish of chillies, the '3' signifies the 3rd Brigade, the '5' the fifth pilot with the unit. Perhaps Tamale was an apt name for someone flying a vehicle that rained down fire and destruction. I had been given rough map-grid coordinates for the position of the troops-in-contact with the enemy before I had taken off. I pulled out a miniature torch from my flying suit pocket to read the map, and with the aid of occasional flashes of moonlight I set a course for the contact area.

I was always a little apprehensive because of undisciplined artillery fired indiscriminately by soldiers of the Army of the Republic of Vietnam (ARVN) into the general area whenever there is a report of allied troops in trouble. In any case the ARVN was riddled with spies. The Viet Cong also monitored our radio communications and frequently interfered with our radio reception by using jamming transmissions. I was told that the VC had files on all the FACs and this was frequently confirmed for me when they kept transmitting in Vietnamese '*Upt-dai-loi, die wie*', meaning, 'Australian captain'. All this added another level of complexity to my job.

It was now fifteen minutes since I took off and I was near the site of the action. I called the commander of the soldiers on the ground.

'Big Trooper 66, this is Tamale 35. Are you having some fun?'

A soft whispered voice came back, 'Roger that, Tamale'. He continued, 'We have taken several casualties and are pinned down. How long before we can get some TACAIR [Tactical Air or fighters]?'

'About fifteen minutes, Big Trooper 66', I told him. 'I have Moonshine [flare illumination aircraft] on the way but I will not light up until the fighters are ready to go. Stay low and I will get back to you soon.'

The enemy below, who had previously held the advantage through their surprise assault on the US Infantry, would have begun to feel a little uneasy with the arrival of my light aircraft overhead. They would have heard it somewhere in the darkness above them. They knew that to fire on the Birddog would almost certainly bring a swift and brutal retaliation if they do not bring it down with their first burst of fire. So they would have held their fire.

Suddenly in my headset crackled another voice on the UHF radio.

'Hello, Tamale, this is Moonshine 74. You have a ball game going?' The call was flippant and typical. Perhaps the American's bravado was a conscious mask to conceal any fear that might have been detected in his voice. I respond.

'Affirmative, Moonshine. This is Tamale 35. We have troops-in-contact and they have taken casualties. There is no ground fire as yet. Your rendezvous will be 200 at 62 off channel 73.' This told the approaching flare ship that channel 73 was the frequency of the Tactical Air Navigation (TACAN) beacon he should use to find the contact. The 200 was the bearing he should use from the station for his heading to the scene of the action, and 62 miles from that point. The FAC was not only a pilot and a battlefield commander, but also an accurate and fast mathematician.

The flare ship was a Second World War-vintage Douglas C47 Dakota known to many civilians as the DC3. It was a twin-propeller cargo plane specially fitted out for its role in Vietnam. Its call-sign, 'Moonshine', hinted to the aircraft's responsibility. As long as it kept dropping flares, it could turn the immediate area around the battlefield from night into day. The flares were one million candlepower and they burnt from one-and-a-half to two minutes as they slowly descend, aided by a small parachute. The flare-ship commander needed to allow for varying wind drift at several levels during the descent.

The flare-ship pilot came back to me, 'Roger, Tamale. I will be with you in ten minutes.'

My three radios had started to get busy. FM radio was used to speak with the troops on the ground. UHF was used to communicate with other aircraft. VHF was used to speak with my TACP, and other more distant bases and aircraft.

The enemy below was keeping a low profile, as they wouldn't have wanted me to know where they were. I call my TACP.

'Tamale Control, this is Tamale 35. Have you any word of my fighters?' This question made it clear to all on the radio net who was now in charge.

'Roger that, 35. You will be getting 'Bobcat 51' in five minutes, followed by 'Buzzard 31'; both carrying wall-to-wall 750 high-drags and 20 mike-mike. Will you require more fighters than this?'

'Control this is 35, not at this stage. We will see how the action goes.'

The curious descriptions were emphatic and ubiquitous. The USAF radio operator back at base had reported that the approaching fighters would be armed with 750-pound (340 kg) high-explosive high-drag bombs and 20 mm calibre cannons. These bombs were dropped by the aircraft on a horizontal run at a height of 20 to 50 metres. As the bombs were released from the aircraft, fins were deployed from their rear. The drag of the fins through the air slowed the descent and increased their accuracy. High-drag bombs could be delivered with much greater precision than conventional bombs released in a steep dive from a higher altitude. The 20 mm cannons were used for strafing runs after all the bombs had been used. I knew from previous operations what type of ordnance is required for a specific task. In a situation where friendly troops were close to the enemy, maximum accuracy was required with no room for error. I also knew from the radio call-signs that the aircraft coming would be F100 Super Sabres, colloquially known as 'Huns'. The F100 was a fine aircraft. There were many hundreds operating throughout Vietnam. They were versatile and accurate fighter-bombers flown by first-class pilots.

As the fighter-bombers were on the way I called the infantry commander below to explain what was going to take place. From my earlier conversation with the ground commander I had been able to roughly estimate where both the friendly troops and the Viet Cong were located. I would finetune this when the AC47 flare-ship arrived.

'Big Trooper 66, this is Tamale 35, we will be dropping flares in five minutes and laying down ordnance two minutes later. Have your grunts keep their heads down and no peeping, as we will be bombing within 150 metres of your position.'

'Roger, Tamale. Understood.' Big Trooper sounded more relaxed. The chatter on the radio networks, the activity on the ground below and in the air space above, all intensified as the two Super Sabres approached.

'Hello, Tamale, this is Bobcat 51.'

'Hello, Bobcat 51. We have troops-in-contact and have taken

casualties. Your rendezvous will be 200 at 62 off channel 73', I told the fighter pilot.

'Roger, Tamale. 200 at 62 off channel 73. We are two F100s, mission number 6152 carrying eight 750 high-drags and 20 mike-mike. We are approaching the rendezvous, switch off your nav-lights. I think I have you visual.'

The 'nav-lights' were my navigation lights. I wrote all the fighter's details on my windscreen with a black chinagraph pencil for reference then and later when I filled out my mission report. The mission number was given so that the fighter's call-signs could be logged and linked to this mission. There were hundreds of other missions carried out each day along the length and breadth of Vietnam. It also provided my TACP and myself with a reference number so that bomb damage assessment (BDA) could be forwarded to the fighter unit at the completion of the mission. It was a mundane and bureaucratic aspect of what could be a high-drama, adrenalin-pumping adventure for the fighters. Also, fighter pilots needed to know how they performed during the mission. It was good for morale to know when you had done a good job.

'Bobcat, this is Tamale, I have you directly overhead. Are you ready for target briefing?'

'We have you, Tamale. Go ahead.'

'Bobcat 51, we have troops-in-contact with VC in bunkers along the canal directly below me. Co-ords XS 036 458. The 2000-foot winds are 110 at 20, surface winds at 090 at 10, altimeter setting is 29.78. Elevation of the target including trees is 60 feet [18 m]. The friendlies are 150 metres east of the target. I expect some ground fire. Your best bailout will be six klicks east over that nearest group of lights. In the target area bail out within one mile [1.6 km] of the friendlies. Closest emergency airfield is Binh Thuy, channel 112. Your attack will be along the east bank of the canal below me running in from southeast to northwest and breaking southwest. Moonshine will be running a reciprocal pattern at 5500 feet [1700 m] and breaking northeast. The FAC will be from zero to 1500 feet [460 m] over the target.'

There was a pause between each of these instructions so that the fighter pilot could assimilate the information. I had continued to add information to my side windscreen with the chinagraph pencil. The Cessna's windscreen was my mission notepad and *aide-mémoire*.

'Tamale, this is Bobcat 51. Attack southeast to northwest, friendlies 150 metres east, bailout six klicks east, 29.78.' For brevity the pilot only read back the most important items.

I had told the fighter pilots of the wind direction and speed at both 2000 feet and on the ground to help the F100s line up with their run-in, and also to determine at what point to drop their bombs. The reference to the altimeter setting was another parameter for the pilot's bomb-release point. The elevation reference of the target was crucial to the height at which the bombs were dropped and how the aircraft climbed up and out after the release. I had also assured the pilots that there were no mountains in the area that might prematurely and catastrophically end their mission and advised them of emergency measures should they be hit and disabled by ground fire. My knowledge of the closest emergency airstrip and its beacon frequency had come from my numerous flying missions throughout the Mekong Delta area of South Vietnam. I had advised the pilots of the direction of their attack and the direction they should break as they pull away from the action. Lastly, I had outlined where the flare ship would be. All of this was a specific form of combat shorthand. It helped the pilots to drop their ordnance exactly where I wanted it. In reality, this was my mission and my command and direction of it would determine its success or failure.

On some targets, where there were no friendly forces to consider, I could give the fighter pilots the benefit of random-attack headings, which made them less vulnerable to ground fire. Flying the same attack heading each pass gave the enemy an advantage—they would know just where the fighters were coming from. All they had to do was fire into the air, without sighting, when they heard the aircraft approaching. Random headings relieved this problem but increased the workload of the FAC as he had to be vigilant to anticipate where the fighters were at all times. This was difficult enough by day but at night it was very demanding and a number of aircraft had ended up colliding during night operations.

Having briefed the fighters I now addressed the flare ship.

'Moonshine, this is Tamale. Copy the fighter briefing?'

'Roger that! We will be ready to drop in one minute.'

'OK, Moonshine. Drop when ready.'

After checking that the flare ship had heard and understood the fighter briefing, I was then busy talking to five stations on my three radios as the first movement of this deadly symphony began.

'Bobcat, this is Tamale. Set up your pattern. I am directly over the target now.'

Just then the second set of fighters checked in.

'Hello, Tamale, this is Buzzard 31. We copied the briefing and will be with you in five minutes.'

This news was great to me as it would save doing a full second briefing and I could run Buzzard in straight after Bobcat with hardly a pause. I vectored the new arrivals in on the same radial as the first pair of attack aircraft. Moonshine 74 was now dropping flares as he slowly circled above the battlefield. I was watching from the Cessna and I needed to adjust the flare ship's drop pattern.

'That was a reasonable flare, Moonshine. Move the next one 500 metres further west. How did it suit you, Bobcat?'

'OK, Tamale, 500 metres west will be real fine.'

The intense glow from the slowly descending parachute flares lit up the immediate area in a ghostly glow and I could then see both the US infantry and the VC troops scurrying for cover. However, the illumination was not quite where I wanted it, so I called for another adjustment. With the increase in activity, some of the previous formality disappeared from the radio procedure and the chatter was almost like several telephone conversations taking place simultaneously.

'Bobcat, this is Tamale. We are ready to go to work. Call when you want the target marked.'

FLASHBACK

'Give me 30 seconds to get into position, Tamale.'

'Moonshine, this is Tamale, start putting out those flares two at a time.'

'Roger, Tamale.'

Like a race horse on a training circuit the C47 flare ship constantly circled the location, dropping a steady stream of parachute flares that drifted downward and brought light to the previous terror of the American infantryman's darkness. Now I joined the action again and sent out a general call. From 500 feet (150 m) I stood the O-1 in a climbing, 90-degree wingover towards the target and as I dived I transmitted, 'FAC going in for the mark. I will put my mark at the eastern edge of that patch of nipa-palm 20 metres from the canal bank.'

I was now the lead actor in this show and they all focussed their attention on my little aircraft. I was about to test the VC, who had been keeping a low profile. If I marked close to their position they would open up with all they have. I dived straight at an area where I'd seen VC scurrying in the light of the first flare. The altimeter spun crazily and at 100 feet (30 m) there was a sudden jolt and a brief flurry of sparks as a marker rocket ignited and sped down and forward off the wing of the Birddog. The rocket sighting system was rudimentary. It was nothing more high-tech than lining up the target with the Chinagraph pencil mark I had drawn on the windscreen. From practice, I'd found it a good sighting reference. I had previously zeroed this cross on the windscreen so I'd known that the rocket would go where I aimed. For accuracy it was essential that the rocket was fired as close as possible to the target, and therein laid the danger.

My marker hit the target and white smoke, clearly visible in the flare-lit night, billowed out. I pulled back on the stick and slowly, agonisingly the Birddog shuddered as it climbed out of this dangerous area. Like a disturbed wasp's nest the enemy fire was directed at the climbing Cessna. The VC knew that if they could shoot down my aircraft, then it would not be possible for the fighter aircraft, waiting overhead, to continue the attack. The pungent odour of cordite mixed with disturbed swamp was evident in the cockpit as it was sucked in through the open window on the starboard side. I could see tracer fire fanning out and I also noted a number of helicopter gunships arriving like zealous pit bull terriers, anxious to join the battle. Right then I didn't want them there.

As I was pulling off the target I angrily transmitted on FM, 'Big Trooper 66, this is Tamale. Get your choppers out of here and have them hold five klicks east of your position.'

'Will do Tamale', was the apologetic reply.

I again focussed on the job at hand. 'Bobcat, I have your contact. Put your first bombs ten metres west of my mark', I told the approaching fighter. He was now clear for a bombing run onto the illuminated target. 'Bobcat Lead, you're cleared hot, lots of small calibre ground fire, I will hold 200 metres west to draw their fire.'

'Roger that. Thanks, Tamale. I have you and I have your mark,' radioed Bobcat Lead.

The fighter flashed underneath the O-1 and the roar of the jet engine was distinct and unmistakable. A few seconds later there was a bright orange flash and a loud explosion as the two high-drag 750-pound bombs detonated. The small O-1 bucked in the shock wave, as I was less than 200 feet (60 metres) from the detonation. Major Nelson in the back seat was on the intercom and agitated. I didn't have time to listen to him. This experience was all new to him. The steady streams of tracer fire lacing up towards us wasn't helping him to feel comfortable either. I must say that I had not been doing much training up to this point as I had virtually forgotten he was there. My whole being was totally involved in the action.

The ordnance had overshot my mark by ten metres. I adjusted the fighter's strike point as the lead Super Sabre pulled out and his wingman made his approach. I was never satisfied with 'close enough', I wanted the bombs on target.

'Lead, that was ten metres at twelve o'clock. Two [referring to Bobcat 2], put your bombs ten metres west of the base of my smoke. Cleared hot.'

I maintained control over the mission while adjusting the target point for the second fighter. By directing the fighter to aim for the base of my smoke I had offset any chance that the pilot might have aimed for any other point on the now drifting smoke.

'Roger, Tamale. I have you and I have the target, in hot.'

Again my aircraft bounced about in the turbulence caused by the bomb's detonation.

'A good bomb there, Two—direct hit. Lead, put your next bomb 30 metres east of my smoke.' I was creating a spread pattern to increase the chances of hitting any nearby bunkers or fleeing Viet Cong. I intended to saturate the area with 750-pound bombs.

'OK, Tamale. I have you and the smoke, am I clear?'

'Cleared hot, Lead.'

The next drop was a good one and it had had an effect as the enemy tracer fire directed at my aircraft intensified. Tracers were arcing up towards the F100s as well.

'Bobcat, this is Tamale. I am getting heavy fire from a point closer to the canal bank. Standby and I will mark it!'

Despite protests from the back seat, I lined the aircraft up for the next rocket strike. I think Major Nelson was feeling rather vulnerable and wishing he was elsewhere. This was his first mission since he arrived in Vietnam and it was probably not a good idea to bring a first-timer along, but he'd been keen to learn and insisted on joining me. I later learnt that Major Nelson did not agree with the war in Vietnam but that did not prevent him from doing an outstanding job for his country.

I rolled the Birddog into a wingover and made a beeline for the target. With the blinding flares alternating with asymmetrical shadows, no visible horizon, carrying out aerobatic manoeuvres while looking over one's shoulder was a pilot's worst instrument-flying nightmare. After flying in these conditions, I never had trouble with instrument flying ever again. At 200 feet (60 metres) the rocket thudded away in a shower of sparks. I could hear the heavy automatic fire from the cockpit as the O-1 struggled to climb out of danger. Adding to the pyrotechnic display I could see the exchange of tracer fire between the now more confident US force and the beleaguered enemy as the tables began to turn. The little aircraft pulled itself up and my head set crackled, 'Tamale, this is Big Trooper—an RPG...'

That was all I heard before there was a simultaneous deafening explosion close by that threw the Cessna up on its nose and starboard wingtip, almost onto its back. The effects of the explosion were probably amplified by my involuntary tugging at the controls with the shock of the event. Instinctively I felt for the lucky charm that I always wore when I flew. It was not there. I fought to gain control of the plane while I waged a battle with a momentary panic that threatened to overwhelm me. As I looked back I saw my trainee, wide-eyed and very pale. At the same time as I heard the detonation of the ground-launched rocket propelled grenade (RPG) I'd felt and heard the impact of shrapnel cutting into the Birddog. But the gallant little craft flew on. There was no time to worry about the damage. It was still flying and that's all that mattered. If it fell apart later, then so be it. My troops still needed help and, as I was unceremoniously hanging upside down, I could clearly see the smoke trail of the RPG. It identified the launch point of the rocket. Now anger had taken over from my panic and, from my inverted position, I only had to make a small adjustment to the O-1's attitude to send a Willy Pete back along the smoke trail of the RPG, directly towards the launch point.

'Crump!' My marker rocket sped away towards the target. Due to the delay caused by my unscheduled manoeuvres, Bobcat 2 had gone through dry but Bobcat 1 was just rolling in calling, 'Christ! That was interesting. Lead in hot. Which target, Tamale?'

'Bobcat Lead—Hit my last smoke!'

All of this happened almost simultaneously and there would have been no more than a minute between the enemy launching their RPG and them being hit by the fighter's bombs. The VC who had fired the rocket had but a few seconds to reflect on their folly before they were sent to oblivion by a pair of 750-pound bombs from the thundering Super Sabre.

There was a large secondary explosion that meant the fighter has received a bonus by hitting either a stock of RPGs or an ammunition dump. I was vengefully delighted with the outcome. The US pilot in the back seat seemed astounded at what he took for the cool actions of his mentor. Little did he know that I was shaking all over with fear and it was taking a lot of effort to make my voice sound calm and casual.

I took my aircraft back 600 feet (180 metres). The fighters screamed in under me and released their bombs. The lead fighter had used up his bombs and his next go would be a strafing run, using 20 mm cannons. The Super Sabres have four 20 mm cannons —under the fuselage. Each gun has 1500 rounds of ammunition. The Bobcats screamed in one after the other in strafing runs parallel to the canal bank, 100 feet (60 metres) off the deck. During the last pass Bobcat Lead was hit by ground fire. I could hear his heavy breathing over the radio as he inhaled oxygen through his face mask.

'Tamale, this is Bobcat Lead. I have been hit and I am losing hydraulic oil. I am heading home.'

'Roger, Bobcat 51. Good luck and have a safe trip. That was great bombing. I will send your BDA [bomb damage assessment] through the TOC [Tactical Operations Centre],' I replied.

'Take care, Tamale', radioed Bobcat Lead as he departed into the darkness. He was probably wondering how Tamale would fare for the rest of the night.

The back-seat trainee, himself an experienced USAF jet pilot and senior officer, was obviously impressed by my handling of the operation. However, he was having trouble speaking without stammering. I was far from relaxed myself, with the adrenalin pumping, but despite this, and the departure of the first set of fighters, the mission went on.

I called the fighters I had loitering overhead.

'Buzzard 31, this is Tamale 35. You ready to go to work?'

'Yeah, Tamale. We are two F100s, mission number 3151, carrying four 500-pound [230 kg] slicks, four un-finned napes, 20 mike-mike and we have the target—looks a bit hot down there!'

The ordnance of the new fighters was different from that of the last F100 Super Sabres. The 500-pound slicks were conventional iron bombs and they were dropped at a higher elevation of about 2500 feet (760 metre) as the fighters dived on the target at a steep angle. They were not as accurate as the low-release high-drags. The new aircraft were also carrying napalm in canisters not fitted with guidance fins. The VC must have dreaded this ordnance, which spreads and consumes oxygen from the surrounding air as it burns. But it was also unpredictable in how far it might travel once it hit the ground and ignited. It was effective but rarely used in close-in troops-in-contact situations. In this case the fighters would have to make their runs parallel to the line of the friendly ground troops to avoid spreading the burning napalm on their own troops.

I didn't want to dwell on the matter and ignored Buzzard Lead's comment about it being 'hot down there'. This was all a part of my act of nonchalance! 'OK, Buzzard, come on down. You will be going to work in one minute. Moonshine, those flares need to go 200 metres west.'

'Roger that, Tamale.'

'Big Trooper 66, this is Tamale 35. The next fighters will be at work within one minute and I have two more sets on the way', I told the infantry commander below.

'That's real fine, Tamale. Those last bombs caused Charlie to leave the tree line and we picked a few off.'

The enemy troops were beginning to break and scatter.

'Buzzard, this is Tamale, confirming friendlies 150 metres east. Attack southeast to northwest and break southwest. Moonshine is running reciprocals at 5500 feet [1700 m] with the FAC over the target from zero to 500 feet [150 m].'

'Roger, Tamale. Friendlies 150 metres east. Attack southeast to northwest and break southwest.'

The VC had stopped firing and were probably on the move out of the area. There was only one line of withdrawal and that was along the sides of the canal. Again, I rolled the O-1 over and made a shallow dive in behind some trees. I made the Birddog weave about like a drunk but drew no significant ground fire from the

VC. I decided to saturate the probable lines of retreat of the enemy. I pulled the aircraft up steeply and performed a wingover through 90 degrees calling, 'FAC is in for the mark'.

I pulled the nose down below the horizontal, rolling wings level and I put my smoke marker down in a likely area ahead of the retreat. 'Buzzard Lead, let's take your slicks in salvo. You're clear hot. Hit my smoke!'

'Roger, Tamale. I have you and the smoke. In hot.'

I had decided to use the 500-pound slicks in salvo—that is, all in one pass—as I wanted to spread the napalm around and did not consider the slicks desirable ordnance for this type of work. This time the fighter released from about 2500 feet (760 m) and I was able to observe the bombs leave his aircraft through the top of my windscreen and follow them down to impact only a few metres in front of me. I was a little apprehensive for a while as I felt I may have misjudged the bomb's trajectory due to their proximity as they sailed past me. The explosion of the four 500-pound bombs was a little scary as I was only at 500 feet (150 m) and my aircraft was severely jolted. I reminded myself to stand off further for Buzzard 2's pass.

'OK, Buzzard 2. Salvo your slicks 50 metres north of Lead's impact.' Lead's bomb impact was still obvious from the burning timber.

'Roger, Tamale. Buzzard 2 is in hot.'

'Buzzard 2, you're cleared hot.' This time I did not observe the bombs leave the F100 as I was turning to position for the next pass, so I was trusting that Buzzard 2 did not deviate from the last attitude I had seen him in. Ka-boom! The bombs landed right where I wanted them.

No sooner had the bombs exploded than I was rolling in for another pass on a different area I had selected, calling, 'FAC is in for the mark, Buzzard. Let's use napalm this time.'

'Roger, Tamale.'

This time I was concentrating on an area 100 metres further northeast. Away went my Willy Pete and it blossomed nicely, indicating good solid ground for the napalm to spread out on.

Buzzard Lead had tightened his pattern and called apprehensively, 'Buzzard Lead, in hot with nape, am I clear?'

'Buzzard Lead, cleared hot.'

Napalm was released in level flight at no more than 100 feet (30 m) and I barely had enough time to clear the target when the napalm started igniting. I was far too close for comfort, as I felt the heat and concussion, and I could smell the burning napalm. I must have been getting tired, or careless, or perhaps both!

'Right where I wanted it, Lead. Buzzard 2, lay your nape down parallel and east of Lead's.'

'Roger, Tamale. Buzzard 2, in hot.'

'Cleared hot, 2.'

Once again there was an enormous flash and an advancing wall of flame as the napalm spread.

'OK, Buzzard, let's expend your 20 mike-mike up the east side of your last nape. Kick the rudders around for spread and lay it down for a klick in length. I will keep out of your way. Buzzard Lead, cleared hot.'

'OK, Tamale. Buzzard Lead, 20 mike-mike, in hot.'

Clearly audible through the open window was the ratter-tat-tat of the guns as the 20 mm cannons chewed up everything in their path.

'Buzzard 2 is in hot, 20 mike-mike.'

'Buzzard 2, cleared hot.'

After Buzzard 2 had completed his run I cleared both aircraft to leave the area. 'Nice work, Buzzard 31. I will send a BDA through the TOC when I have had a chance to look over the area later today and the Grunts send in their report.'

'Great working with you, Tamale. Call by the hooch next time you're in Bien Hoa. I'll buy you a beer.'

'Keep them cold, but not Ba Ma Ba!' I said, continuing the bravado. Ba Ma Ba was '33' Vietnamese Beer, also known as 'Tiger Piss' because it was then a disgusting brew.

'Roger that, and save a beer for me,' transmitted Moonshine. I had almost forgotten the flare ship, as they had been so efficient and they'd hardly required my attention for any readjustment of the flares.

'That was well done, Moonshine. No further illumination required. I plan to put in arty now so you are cleared from the area A-sap. Say hello to Mum!'

'Thanks, Tamale. A nice job.'

I felt the compliments and bravado that fly around during combat brought out the best performance from men. To receive criticism at such times of extreme stress would be quite shattering to one's ego and performance.

By the time the fighters left the ground fire had ceased and the enemy were too widely dispersed for the effective use of air strikes. However, I was not finished.

'Tamale Control, this is Tamale 35. We will not need any further fighters. Give me the call-sign and frequency of the nearest artillery.'

'Roger 35, I can sure use those extra fighters I ordered with Tamale 33 who has a contact south of your position. Stand-by for the arty freq.'

As I slowly circled the scene below I searched the gloom, seeking targets of opportunity.

'Big Trooper 66, this is Tamale. That is all the TACAIR I have for you but I have seen some movement in the open paddies on the west bank of the canal so I intend to put a few rounds of VT fusing on them. I will be off fox-mike while I am doing this.' I only had one FM radio.

'OK, Tamale. We have not moved so the movement must be VC. Let me know when you're finished.'

The VT fusing referred to variable time fused artillery rounds. They could be set to explode 200 feet (60 metres) above the ground, showering an acre (less than half a hectare) with high-velocity shrapnel. It was a deadly and effective rain of hot steel. However, the fuse could sometimes be triggered by flocks of birds in flight or even foliage, so it was prudent to keep well away from their trajectories. The ever-vigilant radio operator back at the TOCP did not cut in until I had finished with Buzzard, which indicated his professional knowledge of operations. Immediately I had finished with the fighters on Uniform, 'Tamale 35, this is

Control,' came in on Victor. Victor was my very high frequency (VHF) radio.

'Go ahead, Control.'

'Your nearest arty is "Slingshot" at Vinh Long, 89.9 fox-mike.'

I dialled up 89.9 on my FM radio and call, 'Slingshot, this is Tamale 35.' My efficient TOC has already alerted them as they came back instantly.

'Tamale 35, go.'

'Slingshot, I require VT fusing for TIC, co-ods XS 036 458.'

'Roger, Tamale. Standby for a marker.'

Seconds later a white phosphorous puff of smoke appeared from nowhere near the coordinates I had given. It was almost incandescent and clearly visible in the darkened skies. The fall was not exactly where I'd wanted it, but it was well away from the friendlies and we were going for spread anyway. Also, I had not wanted to give the VC time to take cover as the marker round was unfortunately a warning for them as to what was about to happen. So I immediately transmitted, 'Slingshot, twelve rounds of VT, fire for effect.'

I had already positioned myself north of the target on the opposite side from the approaching artillery. For the next few minutes the open area and any vegetation in it was literally chewed up by the high-velocity shrapnel raining down from the VT fusing. As soon as the barrage stopped I heard, 'How's that, Tamale?'

'Standby, Slingshot. I will go in for a look.'

By this time the dust had settled and objects were quite discernable on the ground in the early dawn light. I descended to zero feet and flew around the area eaten up by the artillery. There was no gunfire. The morning air was smooth. The devastation hit me as unbelievable as I could not see what effect our attacks had been having in the dark. Suddenly in the morning gloom the stark reality of it all was revealed. The artillery and the bombardment had devastated the jungle. Nature bore the scars of the conflict. She would recover but many would not.

I radioed the FSB.

'Slingshot, that was well done. I can give you 6 KBA.' I saw six bodies in the open fields.

'Thanks, Tamale. Nice working with you.'

I then flew low over the Grunts and advised, 'Big Trooper, this is Tamale, you can call in your gun-ships now to mop up. That's all I have for you.'

'Thanks a lot, Tamale. That saved our arse. We will catch you back at Dong Tam.'

While I was down low I decided to do a BDA (bomb damage assessment) rather than come back later in the day. I added the BDA to the details on my windscreen for my after action records. I had been on target for three-and-a-half hours. Slowly I climbed out of the contact area into the rising sun and was suddenly overtaken by fatigue.

I spoke to Major Nelson on intercom in the back seat, 'Would you care to fly it back to Dong Tam, Sir?'

He replied, 'No, you appear to have things under control. I am going to take some shut-eye.'

As I looked through the windscreen the scene changed. I was inspecting the aircraft and found that the RPG had damaged the rudder, which would have to be replaced. I later learnt that US infantry clearing patrols had swept through the scene of the battle and found 81 Viet Cong bodies. I did not feel elated about this, nor did I feel remorseful. I peered ahead through the aircraft windscreen and saw the beginnings of another dawn far out at the edge of the horizon.

Back in the Boeing we have left the tropical storms behind. The sky is clear and the stars look like the lights of some giant city in the heavens. To my right I can see Orion starting to set and Penny diligently going about her tasks. How can she look so fresh after a night without sleep! As I jolt back to the present, I become aware of someone standing behind me.

'Would you like a cup of tea with breakfast Captain Cooper?' enquired the flight attendant.

'Yes, Elizabeth. No milk or sugar and poached eggs, please.'

What a contrast this is to 1968 in the Vietnam Delta!

2

THE EARLY YEARS

NOT LONG AFTER I finish my breakfast at 37 000 feet (11 000 metres) the first rays of the morning sun begin spearing upward from the eastern horizon. I still find something magical about this even though my eyes feel as though they are full of sand due to a night of flying without sleep. Ahead we can see patches of cloud but it looks like the east coast of Australia will have a fine day. In another 20 minutes or so I will lower the nose of the aircraft and begin the long descent into a landing pattern that will take us to touch down at Sydney's Kingsford Smith Airport. After passing through Area Control to Approach Control we are finally cleared to call the Tower.

'Sydney Tower, this is Ansett 888 Heavy, ILS Runway 34 left, outer marker, 3000 feet, in cloud', says Penny into her microphone.

'Ansett 888 call when visual,' answers the tower.

'Roger, Ansett 888 Heavy,' Penny responds.

Shortly after this, as we break through the cloud and the familiar Sydney city skyline comes into view, Penny and I confirm we are visual.

'Sydney Tower, Ansett 888, visual runway 34 left,' calls Penny.

THE EARLY YEARS

'Ansett 888, clear to land 34 left.'

'Ansett 888, 34 left,' Penny says as she finishes the exchange with the control tower.

As we approach the runway threshold, Ron calls, '500 feet.' Penny then starts calling down the heights in feet: '100–50–20'. I start easing back on the thrust levers and check the rate of descent with the control wheel—'10'—then 16 tyres kiss the runway in a satisfyingly smooth landing. I pull the four engines into reverse thrust to slow the Boeing 747, which helps to save the brakes wearing. As we taxi off the runway the tower says, 'Ansett 888, call Ground 121.7, and have a happy retirement Captain.'

Bugger! Someone has told them I am finishing, making sure everyone knows. I do not respond but Penny thanks them on my behalf.

We park the aircraft and Penny, Ron, Mark and I are brought a glass of champagne by Elizabeth and we all warmly shake hands. It all seems too final to me. But, the day is not over yet as I have to get Mark and myself onto a domestic flight from Sydney to Ballina in northern NSW. We live on a macadamia farm at Alstonville and Ballina is the nearest airport. We catch a shuttle coach across to the domestic side of the airport to see if we can get on a flight. There is no preferential treatment given to an Ansett captain, in fact the treatment is generally indifferent. Today, we come across a friendly personal services officer, who puts Mark in an economy seat and myself in the cockpit jump seat of a company Boeing 737.

Once again we are in the air, on our way home. I will very much miss flying the big planes on the international routes. As a youth I used to fly light aircraft after I first earned my pilot license. When I was three years old my father joined the RAAF (Royal Australian Air Force). He went to the Second World War and never returned. I really didn't know him. I was all of five years of age when he left. So, it was not because of my father that I took up flying. When I was old enough to have an opinion, people used to ask me what I was going to be when I grew up. It would always cause people to smile when I would say proudly that

I wanted to be a swaggie. The swaggies used to come around knocking on my mother's door asking for food and she would usually give them something. Swagmen came of age during the Great Depression, when many people were unemployed and there was no social security. It was not unusual for some of them in those times to have good academic qualifications. In the 1940s, though, they were more likely vagrants, carrying their swags on poles over their shoulders and living on what they could scrounge. Today, vagrants tend to hang around cities and live out of supermarket trolleys. In 1938 they would walk along the railway tracks and country roads getting casual work and receiving handouts whenever possible. I loved the idea of the freedom they had. If they didn't like the place they happened to be, they could just walk away and leave it. I have always had a great love of freedom and independence, which was probably engendered in my early life roaming around the Adelaide Hills.

I was born at Hutt Street Private Hospital in Adelaide, South Australia, on 21 January 1938. I was christened Gary Gordon Cooper but Gary was soon changed to Garry when my parents discovered the film star with his name spelt Gary. Most people pronounce 'Gary' and 'Garry' the same but pronounce 'Mary' and 'Marry' differently. This is a quirk of the English language. Over the years I became a little sensitive to people laughing and comparing me to the film star, so I started using my second name, Gordon. But, come the space era and Gordon Cooper was launched into space. I then went from being an actor to an astronaut!

I spent my early years at Robe, a fishing village on the southeast coast of South Australia, not far from where the mighty Murray River enters the Southern Ocean. Nowadays there is a monument in Robe built to commemorate the thousands of Chinese who landed there between 1856 and 1858 and walked the 400 kilometres to the new goldfields in Bendigo and Ballarat in Victoria. I don't remember too much about Robe. My mother moved us to Adelaide after my father went to war and we lived at St Leonards and later in other Adelaide suburbs. I had a normal Australian childhood, doing the things that scare mothers, like

THE EARLY YEARS

playing on railway lines and raiding fruit orchards. We would play Cops and Robbers, Cowboys and Indians and the other things kids do. I used to love going with my uncle into the Adelaide Hills to trap rabbits. He and I would go out in the bush for three or four days at a time. We would set the traps in the afternoon and go around the following morning collecting the rabbits. My uncle used to kill the animals by snapping their necks. I was not strong enough to do that and I felt sorry for the rabbits so I used to let a few go. I tried to pretend that they had escaped. I'm sure my uncle knew full well that I had let them go but he said nothing. After we had collected the rabbits we would hang them on a pole alongside a bush track, with a hessian bag over them to keep the blowflies off. The Farmers' Cooperative used to come along with a truck and collect them. My uncle was paid three pence each for the whole rabbit with skin, which was left in the bag on the pole for later collection. Of course, today money, and even trapped rabbits, would not last five minutes left unattended on the side of the road! What a long way society has come along the highway of honesty and integrity!

At times my uncle would skin the rabbits, sell the carcasses to the co-op and then sell the skin separately to the tanners. He could earn an extra penny a rabbit doing this but it was very time consuming. In the evenings we would sometimes have a rabbit cooked on a spit over an open fire with vegetables roasted under the hot embers. The rabbit meat used to taste delicious to me then but I don't much like it today—it's too gamey. Whether the rabbits have changed or my tastes have I don't know. I looked forward to these meals under the night sky back then. On clear nights the stars were awesome in their clarity. I developed a sense of the enormity of space just lying on my back looking up. The isolation and sense of freedom of being in the bush appealed to me immensely.

Perhaps not surprisingly I left school after grade 12 in 1952. Schoolwork could not compare with catching rabbits and the freedom of the bush. My reasoning was that I needed to earn money to help the family; however, my mother was getting by reasonably well. I soon realised that I needed to get a better education so I enrolled with the School of Mines in Adelaide as an external student. I worked all day and took lectures four nights a week for two years to gain my Leaving Honours Certificate. I have always had to study long and hard to qualify as I am a bit dyslexic, which I have had to overcome by repetition and determination. I had no spare time at all in those days. After I received my leaving certificate I then started to study for my pilot's licence and engineering qualification! But my engineering studies were interrupted by job hunting, the time spent in New Guinea, and my subsequent pilot training in the RAAF. So I studied part time until a posting in the RAAF gave me some spare time to take it up again. Consequently, I did not complete the engineering qualification until 1962.

I do not remember having a burning interest in aeroplanes when I left primary school. I probably dismissed the idea of being a pilot, believing it to be beyond my reach and capabilities. My first job was working for a carpenter. I thought then that carpentry was going to be my vocation. At lunch times my colleagues and I would all sit outside to eat lunch and talk about inane subjects like sex and football. After a while I dropped out of these midday conversations and sat looking at the Avro York transport aircraft and Mustang fighters carrying out flying practice over Adelaide. Military flying practice over a populated city would not be allowed today. Watching these aircraft, turning and manoeuvring among the clouds, appealed to my sense of freedom. It was 1952 and all the Second World War heroes were still in the limelight but I could not ever imagine attaining their level. To me they were gods, so I had to content myself with carpentry and listening to the lame lunchtime stories of my workmates with their imagined conquests.

Sometime later, a refrigerator repairman came to our place to fix our fridge. In conversation with my mother while I was standing by he mentioned that he was flying a Tiger Moth on weekends at Parafield, just outside Adelaide. He couldn't help but detect my extreme interest in the aircraft and he offered to take me up with him. So on 10 March 1953, Ray Munn took me for

THE EARLY YEARS

a jaunt in a Tiger Moth. The aircraft's tail number was VH-BWC—I can still remember it. That was my first experience in the air and I was totally blown away by it. From that point on all I wanted to do was to fly. I applied for a job as an apprentice mechanic with the Royal Aero Club of South Australia just to be around aircraft. When I started there my weekly wage was five pounds a week (about $10 in today's money), but unfortunately flying lessons were two pounds twenty pence an hour, so there was no way I could afford to take them.

I heard that there was a sailplane club at Gawler that charged forty pence a launch, usually a five-minute flight. So I rode my pushbike the 40 kilometres there and then back again every weekend to take glider-flying lessons with Brian Creer. Brian later became a renowned aviation consultant and writer. To save riding my bike back on Saturday night I would sleep on the floor of a disused military building on the airfield and ride back home on Sunday evening after spending the day assisting with the launching of gliders. Occasionally my enthusiasm would be rewarded with a gratis five-minute flight. I could normally only afford one five-minute flight each Saturday. Looking back on this I feel I must have been rather dedicated.

In the meantime I joined the Air Training Corps as a cadet and tried for a flying scholarship. Unfortunately I was not successful in gaining sponsorship. It is an irony that none of the young men who received scholarships in my time joined the RAAF. A couple of them joined airlines but most found other occupations outside of flying. However, I was determined to fly full time, but it was not until October 1954 that I could afford to start taking powered flight instruction.

My first flight was with Reg Ellis, who had been a Lancaster pilot during the Second World War. I was in heaven. I could only afford 50 minutes a week but after five hours under instruction, I went solo and gained my Private Pilot Licence 12 months later. To build up flying hours to qualify for a Commercial Pilot Licence, which would enable me to fly for a living, I, and the rest of the budding aviators, would hang around Parafield airport on weekends dressed in Second World War flying boots, jackets and scarves. We made ourselves very visible and conveniently available for members of the public to ask us if we were pilots.

'Well, actually I am, funny you should ask', we would reply. In next to no time we would have them in the passenger seat of a Tiger Moth, Chipmunk or Auster and two pounds worse off in their wallets.

A pharmacist by the name of Bill Treloar from Jamestown, which is north of Adelaide, bought a Ryan STM aircraft. It was a sleek low-wing monoplane that he kept at the aero club. He needed someone to fly it up to Jamestown from Parafield and back each weekend so that he could operate it around country properties over the weekend in conjunction with his chemical business. As I was working at the aero club the chief flying instructor recommended me for the task. I think that Bill Treloar was a little apprehensive at first about entrusting his new aircraft to a 17-year-old; however, after he checked me out I soon became a family member and enjoyed many free flying hours and social weekends at Jamestown. During this introduction to the world of flying at Parafield, I met many young men who would eventually die in aeroplanes. Most of them would die simply by getting their intentions mixed up with their capabilities. I was no better than most and overconfidently took many chances. That I can now look back on this and say to myself, 'You were bloody lucky, Cooper!' has to be a blessing.

Having passed all the academic requirements I finally took my Commercial Pilot Licence test with Eric Eberbach from the Department of Civil Aviation (known as CASA today), but had to wait a month until I turned 19 before the licence was issued. Now I was licenced to kill! Being such a highly experienced ace, or so I thought I was, I could not understand why the airlines were not knocking on my door to offer me employment! After about six months of writing to everyone in the world who had an aircraft, I finally landed a job with Silver City Airways doing flying doctor services into central Australia. This was a rude awakening and I was starting to get a taste of how demanding flying can be.

THE EARLY YEARS

My time with Silver City was not uneventful. On one occasion at Marree I crashed through a fence trying to land in a small paddock alongside the township after dark. Although I had arrived before nightfall, a heavy cloudbank had come over, making it pitch black before it theoretically should have been. In those days we operated without radios so I was very much on my own. The only light I could see in that desert wilderness was from a few houses in the very small town. Someone who had heard me overhead drove to the landing paddock but did not switch off his headlights. I thought he was marking the paddock boundary for me. While landing I did not see the fence until it was too late to abort the landing; however, I must have managed to lift the aircraft slightly as later we could not fit the aircraft between the fence posts to get it back into the larger paddock where I was supposed to have touched down. There were only a few small scratches on the propeller and I mangaged to take off again at first light next day.

On another occasion the Auster backfired with faulty ignition switches while I was trying to start it by hand-swinging the propeller. The engine roared into life and it jumped the chocks. Some very quick footwork by me ensued and I reached into the cockpit and shut down the screaming engine, preventing it becoming a pilot-less machine hurtling across Broken Hill airfield, leaving a trail of destruction behind. After all the commotion settled down and my heart returned to normal function, I noticed I had two broken fingers where the propeller had clipped me on the backfire.

Being rather young and impatient to achieve my goals, time seemed to stand still. The job with Silver City lasted only a couple of months but I was determined to find something else, even though flying jobs were scarce. I caught a bus to Sydney, seeking employment in the big smoke. The only job I could get was as a costing clerk with the National Cash Register Company. My close friend, Graham Wright, also an aviation enthusiast, and I shared an attic room together in Kings Cross above a whorehouse. The window, small as it was, would not close properly so the wind and rain blew in. There was also an almost continuous stream of visitors to the flats below us. Although the girls were very matter of fact with their clients, they were extremely friendly towards Graham and myself. They sometimes fed us and were instrumental in turning a very bleak existence into a tolerable one—and there were no freebies! After what seemed an eternity Graham obtained flying employment in East Africa. He was killed eleven years later on Christmas Day 1968, when he crashed into the side of Longonot volcano in Kenya while on a sightseeing flight. Killed with him was the talented Australian folk singer, Tina Lawnton, who had a voice like silk. It was a tragic loss. They were buried at the site, laying at rest near the inside base of the volcano. Being an adventurer, it is a fitting resting place for Graham.

I was eventually offered a job with Gibbes Sepik Airways (GSA) in New Guinea. The famous 3 Squadron desert pilot Bobby Gibbes DSO, DFC owned this company. I was very lucky to get the job and I suspect my application arrived on a day Gibbes was looking for pilots because, as I found out later, most applications were trashed on receipt, their being too numerous, causing an unnecessary workload on administration. Since I had been with the National Cash Register Company for only a few weeks, I was very apprehensive about resigning. The senior accountant called me into his office along with my immediate superior and wanted to know the reason I was leaving. They were a little relieved when I told them I was taking up a flying career. They said that there had been a lot of people in my job and they were concerned it was reflecting badly on them as managers. Strangely, I thought, they considered I was doing an excellent job and they tried to talk me out of resigning. Personally, I thought I had little idea of what I was doing, I didn't like the job and I was half expecting to be fired for incompetence at any time.

Early in August 1957 I boarded a Qantas Lockheed Constellation from Sydney's Kingsford Smith International Airport. The first stop was Brisbane, where the cockpit crew had to wait in the lounge with the passengers while the aircraft was being refuelled. The crew were talking 'shop' so I moved closer to them

THE EARLY YEARS

to hear their tales about practising flying the Constellation while an engine was shut down. I was in awe. Now, after a lifetime of flying big jets, I realise they were just playing to the admiring public. From Brisbane we flew direct to Port Moresby in New Guinea. I did not sleep during the flight in case I missed something and admired a particular hostess—as we were permitted to call them in those days—mesmerisingly moving up and down the aisle going about her tasks. She must have been in her mid-twenties—far too old for me! Age is relative but at 19, people who were 25 did seem old to me. I did not dare go to sleep in case I missed a glimpse of the cockpit when the hostess opened the door to serve drinks to the 'gods' up front. I thought, 'this is for me', and all my time in New Guinea was spent building experience and dreaming about becoming one of these cockpit gods.

Port Moresby was my first experience outside of Australia. I was excited, happy and enthused about all I saw. The day after I arrived I caught a Qantas DC4 that flew over the Kokoda Trail area to Lae, where I had to stay overnight to get a connection to Goroka, the headquarters of Gibbes Sepik Airways. The steamy jungle, the strange damp smells and scenery fascinated me but my fascination turned to alarm when I noticed the ground at our cruising altitude getting closer and closer as we flew into the central mountain range of the Owen Stanleys on the way to Goroka aboard a Qantas DC3. Also I was getting glimpses through misty cloud of near vertical, jungle-covered mountains on both sides of the aircraft as we flew up the mountain pass. The airport there was about 5000 feet (1500 metres) above sea level. We were cruising at 7000 feet (2100 metres) and the mountains were disappearing into the clouds on both sides of the aircraft. As I peered through the open cockpit door I could see the pilots with their feet up on the forward console, joking and sipping coffee. 'What cool customers', I thought to myself.

The climate at Goroka was vastly different to Lae and Port Moresby due to its height above sea level. The air was much cooler with the nights cold enough to have to wear a coat. Gone was the oppressive humidity of the coast. The mornings would usually be foggy but this would clear by mid-morning to display a beautiful day. By early afternoon tall cumuli piled up and heavy rain would fall into the late evening. Flying conditions could become treacherous very quickly and many pilots were killed by being trapped by the weather in dead-end valleys, flying into the rock face of an obscured mountain.

After two days in Goroka I boarded a GSA Junkers 52 for Wewak on the north coast of New Guinea. Bobby Gibbes had bought three Junkers 52s from the Spanish Air Force, which had been German Luftwaffe paratrooper aircraft during the Second World War. They were well known for their corrugated skin construction. Gibbes had considerable trouble with the original BMW engines so he had them converted to Pratt and Whitney 1340 power plants.

Wewak was a beautiful, scenic, tropical township situated on a hill jutting out into the sea. Most of the houses in Wewak boasted ocean views and cool sea breezes in the afternoons. For a young family it was very pleasant living; however, for a 19-year-old male, there was a distinct lack of the unattached fairer sex. There was one nurse at the local hospital whom I fantasised about, but she was too old for me. She must have been all of 26! In any case, she was involved with one of the other pilots whose name was Doc Wills. Doc Wills was to train me on the Noordyn Norseman, a bulky high-wing single-engine aircraft that was the same type as the one in which Glenn Miller had disappeared over the English Channel in 1944. I cannot remember what Doc Wills' ultimate intentions were in aviation but he subsequently left New Guinea and took a job crop-dusting back in Australia. I heard that he was killed near Armidale in NSW in 1962, when the wing broke off his de Havilland Beaver while in flight. Wewak was also an interesting place because of the large amount of Second World War military equipment still lying around. The next peninsular to Wewak was Wom Point where the Japanese had surrendered to the Allies in 1945.

Being only 19 had its disadvantages. Passengers would do a double take when this young, scrawny kid arrived in a remote mountain area to pick them up. The local natives knew me as

THE EARLY YEARS

'Piccaninny Kaptin', which literally means 'child pilot'. But for a remote location, life flying from Wewak was not too bad. Flying up and down the coast and along the Sepik River was pleasant and the weather conditions not too demanding. I did have some flights into the remote mountain regions around Telefomin, near the then Dutch New Guinea border. The weather there before mid-morning was foggy but by mid afternoon there were mountain-embedding thunderstorms. The airfield at Telefomin had been built during the war as a forward staging point. Gliders carrying small tractors were flown in to construct the airstrip and these aircraft and the equipment were still there in 1958. At the aircraft parking area in Telefomin there is a monument to the two Patrol Officers who were killed by headhunters in 1957.

There was only a small window of opportunity when one could safely operate into this mountain region. In the next valley to Telefomin was Eliptamin, which was only ten minutes flying west along the Eliptamin Valley, through a narrow gorge, then east along the Telefomin Valley. So inhospitable is this area that a Missionary Aviation Fellowship pilot disappeared there in the early 1960s and has not been found to this day. On landing at Telefomin the day before Christmas Day in 1958 I sustained a blow out in my port tyre and radioed Wewak for a spare. There was not sufficient time remaining before the weather closed in so a replacement tyre could not be flown in until the next day. After the heat of the coast it was a pleasant change to sit in front of a log fire sipping coffee with the local missionary that evening. The weather the next day was not very good either but we heard Peter Manser, another GSA pilot, circling further down the valley looking for a gap in the clouds through which to descend. I called him on my aircraft's radio and tried to indicate where the breaks in the weather were but he couldn't get through and had to leave. Pilots in New Guinea generally drank to excess and partied hard, so, being Christmas, this was more of an excuse for them to party on. There was no further attempt to get me out until 3 January, when the enjoyment of and recovery from the Christmas and New Year celebrations were over!

As I wanted to fly multi-engine aircraft, I jumped at the chance when offered a position on the three-engined Junkers 52s based in Madang, which was also on the north coast of New Guinea. Madang was another tropical paradise but, once again, it was lacking in suitable young women. My accommodation was in what had been the military hospital during the Second World War. There were numerous shrapnel holes in the building and the white ants were so bad that the only thing holding it up was the steel frame. Anything made of timber was merely a shell of numerous coats of paint. There was always sawdust on the floor from the termite activities and on still nights you could hear them gnawing away.

Gibbes Sepik Airways attracted a lot of Bobby Gibbes's wartime comrades such as Dick Creswell, 'Blackjack' Walker and Robin Gray. They were all fighter pilots during the war and the company was run more like a fighter squadron than a commercial enterprise. As Bobby Gibbes was spending more and more time setting up and developing his coffee plantation than running the airline, it started to become a bit run down. I was offered a job flying Cessna 170s, 190s and the twin-engine De Havilland Dragon DH84 for Madang Air Services (MAS), out of Madang. I accepted the job as the future at GSA looked rather dismal. This job lasted until early 1960. At that time airline recruiting was in the doldrums and there appeared to be no further possibility of my pursuing a civilian airline career. I spoke with Dick Cresswell about my dilemma and he said, 'Join the RAAF, lad!' I applied and was accepted for interview. The interviews, tests and medicals were conducted over a period of five days in Sydney. On day one there were 400-odd applicants in attendance. This number was reduced each day by elimination and, finally, when I reported to Point Cook for training, there were only five candidates from NSW. During the tests I fell short on the coordination machine as I was trying to fly it like an aeroplane. Squadron Leader Turnnidge, a member of the recruiting panel and the chief flying instructor (CFI) at Advanced Flying Training School (AFTS), took a practical view of this and pushed me through, so I have him to thank for my not being ruled out at that point.

THE EARLY YEARS

After the entrance testing in Sydney there was a waiting period before we knew who had been successful so I went back to New Guinea and flew for Dennis Buchannan's Territory Airlines Limited. Dennis was a young traffic officer with GSA and he had bought a small charter business from the estate of the owner who had been killed in a Tiger Moth in the southern highlands. Dennis was so popular and successful that no one would differentiate between Dennis and GSA—if someone wanted to consign cargo they would say, 'Send it to Dennis!' This had the effect of filling Dennis's aircraft before GSA's aircraft. Gibbes eventually suggested to Dennis that perhaps he should resign from GSA due to a conflict of interests. However, everyone still sent their cargo to Dennis. With GSA floundering, Territory Airlines Limited flourished and soon Dennis was operating several Cessna 180s throughout the highlands. Dennis's Territory Airlines lines eventually went on to become a major internal airline in Papua and New Guinea. After working for Dennis for three short weeks, my acceptance into the RAAF came through. The pilots' course I was to join was to start with the minimum of delay so I was not able to give Dennis the usual two weeks' notice. I told Dennis and he kindly waived the notice period, generously giving me an airline ticket to return to Australia. He certainly was a first-class gentleman. I was very upset to hear several years later that one of his sons commited suicide by jumping from an aircraft over Camden in New South Wales.

I arrived at Point Cook in Victoria on 11 March 1960, the same day as No. 39 Pilots' Course was due to begin. There were 25 hopeful young 'Douglas Baders', who had arrived in dribs and drabs. Most of the cadets were straight out of high school or university, but there were also four re-musters—candidates who had already completed the apprenticeship scheme in the RAAF and done some years service in a trade. There were also a couple of 'older' people, like myself, who had seen a bit of the world and done some flying. We were billeted on the ground floor of a two-storey building, with those doing the course ahead of us, No. 38 Pilots' Course, Basic Flying Training School (BFTS), living upstairs.

Each of us had his own room, which was only two by four metres with a built-in cupboard, a chest of drawers and a bed. No other furniture was allowed, not that one could get anything else into such a small space. It was more a cell than a room. Our ablutions were communal and situated in the centre of the building. The buildings were Second World War-era, and damn cold in winter and hot in summer. The guys doing No. 38 course held a party for us on the first night at the Cadets' Club. The party was really just an excuse for a piss-up. Still, not being a seasoned drinker, I went to bed earlier than most and, as it was hot, I left my window wide open. Bill Clark had the room above mine and at some stage during the night he found it necessary to empty the contents of his stomach out of his window. When I awoke in the morning there were peas, carrots and tomato skins mixed with other bilious gunk on my windowsill and floor. Bill kindly came down and cleaned up the mess for me. The amiable Bill Clark was later killed in 1968 when his Winjeel broke up in flight during aerobatic instruction with an Army student.

The first few days of the course were spent getting clearances, collecting uniforms and generally getting settled in with introductory speeches from all our ground instructors and superiors. We were treated like rookies for the first three months with drill, drill and more drill, sport and physical training—the usual military indoctrination—heavily on the schedule. In the second week we deployed to Point Addis, a bleak, windswept point on the south coast of Victoria. There we lived in tents, arose at 5 o'clock in the morning and ran up and down sand dunes all day. Having not done any real exercise in about seven years, this activity nearly killed me. At the end of the day I would fall onto my stretcher for another cold, uncomfortable night with aching muscles. I was glad when this phase was over.

Back at Point Cook we would do cross-country runs to Laverton and back, which all up was about 15 kilometres. I soon became very fit and actually took second place in the inter-service—Navy, Army and Air Force—cross-country event around the Point Cook Wetlands, a distance of 12 kilometres. There were

THE EARLY YEARS

approximately 60 competitors. About 20 of the entrants tried to lessen the distance by using an old fence cutting through the centre of the wetlands. They got about two-thirds of the way across and found that the fence wires had rusted away, which meant they had to negotiate 200 metres of freezing, knee-deep mud and slush to reach the far bank. We had finished showering by the time these forlorn creatures came staggering back. Poetic justice!

Our Warrant Officer Disciplinary (WOD) and drill sergeant was Warrant Officer 'Ming' Harris. He was barely five feet tall but he carried himself in a very erect, military manner, like he had a broomstick strapped to his back. We used to play tricks on him in retaliation for the belligerent manner he used to treat us with. On drill exercises, we were required to each take it in turn to shout the drill commands and control the squad. There was a technique in throwing one's voice so that it can be heard the full length of the parade ground. Few of us young budding officers managed to master this technique. But when the squad was at the far end of the parade ground, the cadet calling the commands would call, 'Riiiight, Wheel!' and we would pretend not to hear even if we did. We would march straight ahead, off the parade ground, and disappear among the huts with Ming screaming obscenities after us. We would then reappear, still in unbroken rank, along the perimeter road as though nothing had happened. Very much like a rudderless boat. Ming would also spend an hour trying to get us to halt together and we would frustrate him by making each halt sound like an uncoordinated machine gun. Towards the end of hour we showed improvement so Ming would say, 'Right lads, let's show those Headquarters' "poofters" how you can halt!' We would then pass a phrase quietly among ourselves: 'Silent halt!' Ming would march us down the road to the front of HQ and shout, 'Squaaaad, Halt!' We would act to slam our last step down but not make contact with the ground. The sound of 25 pairs of clomping boots suddenly stopping and Ming screeching, 'You bloody bastards', would always bring startled faces to the windows of HQ.

The WOD was issued with a yellow bicycle to get around the base in the course of his duties. Once each month, there would be a big base parade in full uniform, which all units on the base attended. Long Service Medals and the like were presented at these parades. The night before one parade, we borrowed Ming's bicycle and hoisted it up the flagpole. Surprisingly, it was not noticed until the bugle sounded and the flag-party untied the flagpole rope, whereupon a high-speed bicycle came crashing down, narrowly missing the flag-party. Many officers could be seen holding back laughter, even the Group Captain in front of the whole parade had to turn away to hide his mirth. However, their subsequent investigation did not indicate they were amused. We were high on the suspect list but it was never proven that we did it. We could never stop ourselves laughing when Ming rode by on his very out of shape, squeaking bicycle muttering, 'Call yourselves fucking officers!'

After three months of tortuous Rookie activities, we started flying training on the Winjeel. The Winjeel was a low-wing monoplane with side by side seating and a 450 horsepower radial engine. My instructor was Flying Officer Fred Price, who later became a pilot with Trans Australian Airlines (known as TAA) after he left the RAAF. I had already accumulated 2500 hours' flying experience and had little trouble, but Fred was always pushing me to do better. I was surprised how precise and demanding the flying standards were for an *ab initio* school (*ab initio* is Latin for 'from the beginning'). Quite a few of the cadets had to suffer the indignity of running back to dispersal with their parachutes held high above their heads from the far end of the airfield as punishment for some failing in their performance. The instructor would taxi the Winjeel close behind to ensure the cadet did not stop running.

The failure rate was high and by the time we completed BFTS there were only ten of the original 25 cadets left. Flight Lieutenant Alec Young did my final handling test. I passed with flying colours and was awarded the trophy for Most Proficient Pilot. Alec was killed in August 1962 with five other experienced pilots, when the Red Sales Aerobatic Team crashed during a barrel roll just south of East Sale RAAF Base. The men were sorely missed and the accident was a great loss to the air force.

THE EARLY YEARS

The next phase of the pilot training was to be done at the Applied Flying Training School at Pearce in West Australia. We spent the first two weeks receiving lectures on high altitude, high-speed flying and the technical side of the de Havilland Vampire Mk35. On 20 January 1961, I did my first jet flight with me at the controls. It was a marvellous experience! My initial instructor was Flight Lieutenant Max Hayes who seemed very quiet and introverted. We did not get along and he eventually paraded me before the CFI who was still Squadron Leader George Turnnidge. The CFI tactfully explained to me that it was not my fault and allocated me a new instructor, Flying Officer Pat Patterson. Max Hayes went on to do a tour in Vietnam on helicopters and came back suffering from post traumatic stress disorder (PTSD). At the time, few doctors could recognise PTSD symptoms, and he was not treated. He committed suicide, which was a sad ending to a talented man's life.

Pat Patterson and I got along very well, but he would keep pushing me to extend my performance envelope, to get the best out of me. I did not resent this tactic and I tried hard. It is a peculiar feeling to be at the limit of one's mental capacity. One can sense that beyond the barrier there are greater things but you cannot get there. However, little by little you improve as your brain's neurons grow to form more numerous and complex interconnections. Later this habit instilled in me by Pat saved my life on several occasions.

On practicing engine failures after take off, the rule was to never turn back to the airstrip below 200 feet (60 metres) altitude. Pat had me achieving safe landings from a height of 100 feet (30 metres). The resulting manoeuvres were scary but Pat would just sit there with his arms folded as a mark of confidence. Pat Patterson had a short career in the RAAF. He married the daughter of the president of Eastern Airlines in the USA and was the envy of many of his compatriots as he joined the civil flying fraternity in America. He was an excellent teacher.

At Pearce, No. 39 Course lived up to their reputation as being embryo delinquents. This time we had the upstairs floor of the cadet building. As there were only ten of us in 25 rooms, we were able to spread out. Our building was end on to the WRAAF (Woman's RAAF) quarters. From upstairs we could observe the girls going about their chores and passing to and from the showers. We could see that some of them were exhibitionists so we set up a vacant room at their end of our block where we never turned on the lights. We called it the TV room and it made for some pleasant interludes. Geoff Astbury, on 38 Course, had a ground floor room at the WRAAF end and had a false sense of privacy by the high fence between the two buildings. Dave Champion had been receiving treatment for a heat rash on his bottom and dropped his underpants in Geoff's room to enquire how the rash appeared. This was immediately met with catcalls and whistles from the darkened room upstairs in the WRAAF quarters. We then realised that they had a TV room also!

As with our time at Point Cook, we worked hard all week and chased girls on the weekends at Pearce. We were some distance from the city of Perth, so the Course had a two-bedroom apartment rented at Cotteslow Beach. There was only one double bed and three single beds in the apartment, with anywhere up to 15 people staying overnight—not that there was much time for sleeping. When the WRAAFs started using our apartment as well, it really became crowded—and interesting!

On my first solo flight in the Vampire, I had a hydraulic failure. This is not a great problem if one follows the correct emergency procedures. As it was I did all the right things but it was quite a shock to me and my heart was really thumping. We all did dual formation flying with an instructor, but before we could be let loose student-on-student formation flying, we had to be cleared. This meant that one student would be with an instructor while the other would be solo. It was the instructor's job to evaluate both of us. Terry Duggan was to fly with Flight Lieutenant Tex Watson with me solo on the wing. We took off into a low overcast sky and climbed to 33 000 feet (10 000 metres) without breaking clear of cloud. At times the cloud was so thick that I could not see the fuselage of Terry's aircraft. Many times I was so uncomfortable, having only a wing tip and the pitot tube to position myself with. I felt like pulling away, calling lost contact and returning to Pearce by myself. But this would be a mark of failure and I was determined to hang on. The instructor decided the weather was too bad and had Terry carry out a descent using radar, known as Ground Control Approach (GCA), to get below the cloud. On final approach Terry deployed the speed brakes unexpectedly and I overshot him. I turned away and slowed down in accordance with the procedures of lost contact. Well, I thought I had turned away but out of the gloom loomed the sight of a giant wheel in my windscreen. That was all I saw of the other aircraft. Quickly I felt my way back into position on Terry's right wing. During the debriefing Tex Watson said, 'I couldn't see you a lot of the time, Cooper, but noticed you hung in there–good work!' I accepted the compliment but didn't let on how close it had been.

Half-way through our jet training I did an instrument-rating test with George Turnnidge. He just sat there, looking bored, throughout the test and then said, 'Let's get down near the ground and do some aeros!' After I had done some very timid basic manoeuvres at about 1000 feet (300 metres), George took over and carried out some really violent, low-level aerobatics. He was a fighter pilot through and through and no other sort of flying suited him. At the end of my instrument rating he complimented me on my general flying but said I was not showing enough 'dash' in what I did.

As it turned out my final Wings Test was to be done by George Turnnidge also and I was determined to show him a bit of dash. There was a standard procedure for the Wings Test and at the end the student had to do a manoeuvre of his own choice, usually in the vicinity of the airfield, which he was supposed to have practised beforehand. What I planned I could not legally practise. After all the official profiles had been covered, George said, 'Now Cooper, show me what you have in store!'

I dived in a downwind direction at 50 feet (15 metres) above the Pearce runway, at high speed. All the time I was looking out the corner of my eye for a negative response from the CFI.

Nothing, so I continued. I pulled up into a vertical climb and continued rolling until I was low on speed, pulling down into the last half of a loop. At the bottom of the loop I was back over the end of the runway, into wind and about 200 feet (60 metres) off the deck when I pulled up again into a 45-degree climb, commencing a Derry Turn—a roll under—onto down wind, simultaneously dropping the undercarriage and continuing the turn to touchdown. George just sat there with his arms folded. As we taxied off the runway I was beginning to regret my decision to do such a manoeuvre when George shouted, 'Christ, Cooper, now that's dash! Bloody good show!' He had a giant smile on his face, his handlebar mustache was twitching and I could see he had not had so much fun in a long time. It would be two years before I learned what it was like to have the 'fighter pilot disease' as I was initially posted to transport aircraft.

Before I left Pearce, George Turnnidge boldly wrote in my flying logbook, 'Good show, Cooper.' Many people used to ridicule George but I found him to be extremely honest, capable and, most importantly, uncomplicated. In the years to come I was to learn that it seems to be the Australian way to ridicule anyone who is a little different or a cut above the average. The 'tall poppy syndrome' had reared its ugly head in my life for the first time.

At a full-dress parade, the ten successful graduates of No. 39 Pilots' Course received their wings and entered the officers' mess as pilot officers for the first time. I was awarded the Gobel Trophy for most outstanding pilot and Dick Waterfield took the academic trophy as a close second. Dick was a re-muster with the rank of corporal and an extremely good pilot. I was surprised that he did not go on to dizzy heights in the RAAF—perhaps the tall poppy reared again.

On graduation our onward postings were allocated and I drew a staff pilot position with the School of Air Navigation (SAN) at East Sale in Victoria. East Sale was a dismal place. It was extremely cold, suffering from Bass Strait weather. All the buildings were Second World War structures and they lacked insulation, so we would have to run our little two-bar electric heaters all the time.

THE EARLY YEARS

There were cast iron stoves in the flight line crew rooms, which we fired up with wood and heaps of inflammable floor polish to keep warm. The insulation was so bad you could burn yourself on the side facing the stove and be cold on the other side. We would all sit around the heater with our boots up on top of the lid. This eventually dried out and damaged the leather soles and there was the continuous aroma of cooking leather. Severely deformed flying boots would always identify a pilot from East Sale.

At East Sale we flew mainly the C47 Dakota on navigator training exercises, which was rather boring. We would bore around for six hours, sometimes at night and sometimes at 500 feet (150 metres) above the sea, flying headings given to us by trainee navigators. Once each year six Dakotas flew to New Zealand via Amberley in Queensland and Norfolk Island. This was an interesting and enjoyable jaunt. Other less frequent tasks at East Sale were flying the Winjeel, Vampire and Canberra on instrument approach procedures for the Ground Controlled Approach School.

After four months at East Sale, the RAAF called for a volunteer to join the Antarctic Flight for the Summer Tour, attached to the Australia National Antarctic Research Expedition (ANARE). Due to my bush flying and light aircraft background, I became the 'volunteer' and joined Squadron Leader Norm Ashworth, the detachment's commanding officer. Norm had won the Sword of Honour at RAAF College and was quite some academic. I remember first seeing him shortly after he graduated when I was with the Air Training Corps as a cadet, so it was a pleasant surprise to be working with him.

In mid-October 1961 we collected a Beaver aircraft from the de Havilland factory at Bankstown, NSW, and taught ourselves how to fly it. In November we flew the Beaver to Point Cook and fitted floats onto it. About this time two engineers joined us, making a team of four. The two engineers were Sergeant Ron Frecker, an electrical fitter, and Sergeant Alan Richardson, who worked on engines and airframes. Alan had already done a 12-month tour to the Antarctic and was great value as an engineer and a delightful person.

Before leaving for the Antarctic we had to do polar survival training and celestial navigation as flying in the vicinity of the south magnetic pole meant that every direction was north, so special navigational procedures were required. We boarded the icebreaker *Thala Dan*, a Danish ship from Esbjerg built especially for crashing through ice flows and set sail in early December 1961 with the Beaver securely lashed down on the deck of the 5500-ton vessel. The voyage down through the 'roaring forties', 'howling fifties' and the 'screaming sixties' was very uncomfortable with the ship rolling 30 degrees and pitching 20 degrees. It was impossible to remain in bed, even after jamming yourself in. I guess that was why Royal Australian Navy sailors had hammocks to sleep in. One of the scientists worked out mathematically where the centre-of-balance of the ship was in order to situate himself at the point of least movement. It happened to be half way down the ladder into number two hold. He lashed a wickerwork armchair to the ladder and spent a lot of the voyage in it. It was a relief when we got down amongst the pack ice for some smooth sailing.

We spent our whole time in Antarctica on board the ship and only went ashore on brief occasions to weather stations. Being cooped up with three others in a cabin, on a small ship, for three months, was a bit of a strain. It was always enjoyable to get out of the cabin to go flying, even though it was not as simple as jumping in the aircraft and taxiing to the runway. First, we had to get the aircraft into the water. We achieved this by winching the Beaver over the side of the ship, where we would taxi about on the floats looking for an area clear of ice to take-off.

Our task was to carry out tri-met aerial photography (three cameras, one on each side and one vertically mounted on the aircraft to take pictures from horizon to horizon) for the Department of Mapping and to do radar flights inland to ascertain the height of the polar ice. A further duty was supply drops to expedition teams far inland on the ice plateau. On some of these drops we encountered gale-force winds. At these times it was difficult to judge just where to do the drop as the winds would carry the supplies way downwind. We actually had to fly past the teams on the ice with a ground speed of only 20 knots (37 kilometres per hour) or so, and let the parcels blow back. After the first drop it was only necessary to turn the aircraft a few degrees out of wind for the air current to blow us back downwind of the expedition again for the next run in. To do a 360-degree turn would place the aircraft miles downwind, creating a long haul back to the ground team.

Returning to the ship was sometimes perilous, as both Norm and I would have to climb out on top of the icy wings to reach the tips in order to throw the securing ropes to the ship's crew. This was bad enough in calm conditions but it was often very choppy when the plane was on the water, so we were always in fear of being thrown into the sea, where survival was about five minutes. There were two Bell helicopters on board the ship, which doubled as rescue craft if we were forced to land away from the ship due to engine failure. However, the distance we were from the ship most of the time meant that, if we had come down at the limits of our radius while inland, we would have perished in the elements.

We returned to Australia, through the 'screaming sixties', 'howling fifties' and 'roaring forties' in March 1962. After replacing the floats on the Beaver with wheels, Norm and I flew it back to de Havilland's at Bankstown, where the aircraft was stored for next year's expedition. For me, it was back to SAN at East Sale, where my contemporaries were being promoted to captains on Dakota aircraft. It was overlooked that I had been away so I was upgraded along with the others. But after my final check, my flying logbooks had to be endorsed. I was called in and asked by the commanding officer, Squadron Leader Payne, why I was now command training with only 25 flying hours logged on the Dakota when 200 hours were required. I pleaded 'stupidity' with a shrug. A lot of discussion went on at higher levels about this, then Squadron Leader Bartlett, the CO of the Central Flying School (CFS) took me up for another final test. On his imprimatur I was released as a Captain on the Dakota.

I flew with SAN until October when I was summoned by Squadron Leader Payne and advised that I was about to 'volunteer' again for the Antarctic. Normally one does not do two Antarctica expeditions in a row but the CO elect of the expedition, Squadron Leader John Batchelor, was still doing his science degree at Melbourne University. They required an experienced Antarctic pilot to set up the expedition, so as a lowly Pilot Officer, I became the CO of the RAAF Antarctic Flight with an exhaustive team of two sergeants under my command. Alan Richardson joined the Flight again along with Sergeant Don Tiller as the electrical fitter. By the time John Batchelor arrived, I had the expedition set up, with the aircraft on floats and cameras installed and calibrated. All that remained was to get John checked out. This done, John thanked me very much and took command.

December 1962 was spent smashing our way south again through monstrous seas on the icebreaker *Thala Dan* and we started flying early in January 1963. The flying duties were the same as the previous year. However, this time our ship became stuck in the pack ice and we drifted with the ice for over two weeks. We were running out of basic food and had to slaughter seals for meat. Eventually the ice flow broke up as it spread around a peninsular and we commenced flying again.

At Wilkes Station, John and I were invited to visit the Russian base at Vostock, which is close to the South Geomagnetic Pole. Vostock is 11 500 feet (3500 metres) above sea level, sits on 12 200 feet (3700 metres) of ice, and has recorded temperatures of minus 89 degrees Celsius. The place regularly experiences winds in excess of 500 kilometres per hour. We, and some Australian Army personnel, were picked up by a giant Russian Antonov bi-plane and flown the 1200 kilometres inland. On arrival at Vostock, the only visible signs of life were the radio antennas protruding from the expanse of ice. All the structures were buried under the ice. With very little English spoken it was an interesting day. One of the Russian military officers asked an Australian Army officer about the purple and green medal ribbon he was wearing. Without thinking the Australian answered, 'Oh, that was for fighting the commies in Malaya'.

THE EARLY YEARS

All went quiet but we thought the Russians were not sure if they understood and the convivial atmosphere soon returned.

On 19 January I was carrying out a survey of the Vanderford Glacier with two glaciologists, Rex Simon and Alister Battye, when the engine started to run rough. I assumed it was carburettor icing and applied heat, but the engine continued to die. At this time we were over some very uninviting crevasses, among which it would have been a disaster to force-land. I turned for the coast and was fortunate that the slope of the glacier matched the descent rate of the Beaver. We narrowly cleared the ice cliffs and I glided to a landing in the sea off Ivanoff Head. Ironically, we passed right over the wreckage of a TAA helicopter that had crashed on the ice cliffs during an earlier expedition. The three of us used whatever implement we could find to paddle the aircraft to an outcrop of rocks. About an hour later a rescue helicopter arrived with Alan Richardson on board. Alan could not get the engine running properly so we taxied toward the approaching *Thala Dan*, with the engine protesting, backfiring and belching fire and smoke all the way. We were winched aboard and after a carburettor change, we were back flying again the next day.

In March we arrived back in Melbourne but this time Customs allowed us to lower the Beaver over the side of the *Thala Dan* off Point Cook, instead of going into the docks. We still had all the equipment we used in the Antarctic on the aircraft. In the Antarctic we frequently flew at 10 000 feet (3000 metres) with ease but with the density of the Melbourne summer air, we found we could not get the Beaver off the water! Eventually John climbed back on the ship and I dumped most of the fuel from the tanks to lighten the load. This did the trick but I still could not climb higher than 500 feet (150 metres) on the flight back to Point Cook. Soon after we arrived back in Australia I received the sad news that John Batchelor had died on the operating table during a simple kidney stone procedure, due to massive bleeding.

3

THE RAAF AND VIETNAM

I TOOK ACCRUED LEAVE in April 1963, and in the May I reported to 2 Operational Conversion Unit (2 OCU) at RAAF Williamtown in New South Wales to commence fighter pilot training. The F86 Sabre we would eventually fly did not have a two-seat version available and the RAAF was not about to let loose a bunch of untrained pilot officers with lethal weapons in a high-performance aircraft on their own so the first two months were spent learning weapons delivery techniques on the Vampire. On 22 July I did my first Sabre flight in an aircraft with the tail number A94-922. (This aircraft is now being rebuilt in New Zealand to fly at air displays.) My first solo was done with an experienced instructor in a chase plane, flying in formation. On my second solo I experienced a complete electrical failure, which really tested my memory of procedures as I could not talk with anyone on the radio. Flight Lieutenant Bill Monaghan, one of our instructors, was sent up in another Sabre to lead me down to the circuit and we communicated with hand signals until he pulled away for my landing. The fault was caused by a piece of loose locking wire in the electrical compartment becoming lodged across two terminals, shorting out the generator. This was classified as FOD—foreign object damage.

Quite early in training I was detailed to carry out instrument/formation flying with another aircraft, flown by Pilot Officer Ray Butler, on my wing. On crossing the runway threshold the lead pilot calls 'cut' to indicate to the other pilot in formation that he should cut his power and land. The lead aircraft then waits three seconds before he cuts his power and lands. This has the pair nicely placed to decelerate down the runway in formation. Just as I called 'cut' a downpour of rain hit us and when Ray looked ahead he could not see anything. Fortunately he had the gumption to apply power and overshoot as I blew both tyres as I aquaplaned the full length of the runway with very little directional control. After that I tried to avoid mixing the Sabre with wet runways. The fighter training was intensive, but enjoyable, and it continued through to December.

All the members of my course were posted to fighter squadrons at RAAF Butterworth in Malaysia. One half went to 3 Squadron and my half to 77 Squadron. The Commanding Officer was Wing Commander Vic Cannon, DFC, who had flown Spitfires during World War Two and distinguished himself in Korea. The 'A' Flight Commander was the legendary Squadron Leader Bruce Gogerley, DFC, who had shot down the first Russian MIG 15 jet fighter by the RAAF during the Korean War. 'B' and 'C' Flights were commanded by Squadron Leader Jim Kichenside and Flight Lieutenant John Pyman. We were in good hands.

Our time in Malaysia was divided between detachments to Ubon in Thailand, Labuan in Borneo and Singapore. On many of these flights we were armed and we flew on Cambodian and Indonesian border patrols, but we never fired a shot in anger, even though we gave chase to Indonesian insurgent aircraft on many occasions. On one detachment to Ubon I was scrambled with Pilot Officer Jock Bryant on my wing to intercept a radar blip near Nakhon Phanom. We identified it as a Laotian C46 cargo aircraft, with parachute doors open, which was cutting across the southeastern end of Thailand. The radar operator cleared us back to Ubon but later instructed us to shoot the target down. Fortunately we could not comply as we were short on fuel. As it turned out the C46 was in fact authorised and his flight plan had been lost in the system temporarily.

At Butterworth I was detailed to carry out an air-to-air firing range reconnaissance. The purpose was to make sure the range did not have any unauthorised shipping in it prior to running a day of air-to-air firing on an aircraft towed banner. After clearing the range and approaching Penang Island I carried out some low-level aerobatics and inverted flying. Suddenly the main-fire warning light came on so I went through the emergency procedures. Each action on the checklist can lead to the fire warning being extinguished. None worked so I was left with a flamed out engine and not enough altitude to get back to Butterworth. Just as I was getting set to eject over the sea snake-infested water, I caught a glimpse of Bayan Lepas runway, which was the civilian airport on Penang Island. Fortunately there was no civil traffic and I had just enough height to turn base leg and land. A very irate air traffic controller drove up in his car and demanded an explanation as to why I had landed on his airport without authority. After I explained my predicament he was very helpful. The problem proved to be a piece of locking wire that fell from the floor when I was inverted. It became attached to a terminal behind the instrument panel and completed the electrical circuit to the fire warning light. FOD again!

We did a lot of live firing and bombing on various military firing ranges around Malaysia and became very accurate at delivering our ordnance. On one sortie four aircraft obtained 230 rounds on the air-to-air banner out of 400 fired. Mick Feiss landed a 60 per cent score, Jack Ellis 54 per cent, Dave Rogers 42 per cent and I achieved 74 per cent. This was a squadron record. During another low-level firing sortie, Mick Feiss was leading a formation of four aircraft near Singapore. As we pulled up to roll in on the target, Mick saw a Russian submarine sitting just outside the firing range, obviously monitoring our activities. Many of the sorties involved mock dogfights with Royal Air Force and United States Air Force fighters. Our training must have been good because we always managed to get behind our English and American friends and take film of them through our gun-sights.

In December 1965, after two years in Malaysia, Tony Karpys and I were posted back to 2 OCU to complete a Mirage course. Tony was later killed at Williamtown while practicing low-level aerobatics for an air display. The Mirage was a Mach 2 (twice the speed of sound), French-built fighter that was just being introduced to the RAAF. Once again, a two-seat version did not exist at the time so it was a case of 'jump in and go', with a chase plane on our wing. The first few take-offs were done without the use of the after-burner to keep the rate of acceleration down. There were some teething problems and ten per cent of the aircraft were lost in the first three years due to mechanical problems. Nearly all the early pilots had some problems with the aircraft, which tested their mettle.

The Mirage training lasted for three months, at the end of which, I was posted to 75 Squadron. The CO was Wing Commander Jim Flemming, who was a highly experienced Korean War veteran. Jim was a great leader and he had an enthralling personality. I had only been with the squadron for a month when I was scheduled for a 15-second radar trail of Jim Flemming's aircraft. Shortly after take-off, while I was accelerating through 400 knots (740 kilometres per hour) and climbing at 1500 feet (500 metres), the engine gave a loud bang and stopped. I quickly went through the relight procedures, pulling up to trade speed for height. As I reached about 4000 feet (1200 metres) with my speed at the best gliding rate of 240 knots (445 kilometres per hour), I noticed the disused wartime runway of Hexham off to my left. As I continued to try to get the engine running, I turned towards the abandoned airstrip. In my efforts to get the engine restarted while turning away from the village of Raymond Terrace beneath me, I left it too late to eject and was obliged to drop the undercarriage and land. As I was approaching Hexham, Jim Flemming, responding to my 'Mayday' transmission, arrived back and followed me in. I was conscious of the form of Jim's Mirage passing over me as I stood hard on the brakes. The airstrip was only 4300 feet (1300 metres) long and studded with bushes. My touchdown was so hard and my braking so heavy, I had no trouble shedding the

240 knots (440 kilometres per hour) I had been travelling at a few seconds before.

The post-flight inspection revealed an eagle with a two-metre wingspan enmeshed in my engine. I was reprimanded for the force landing as RAAF command claimed it set a bad example to less experienced pilots. Their argument was that it was dangerous because they might have lost a pilot as well as the aircraft. It was basically air force law that a pilot abandoned a flamed out Mirage; that is, they should eject. I argued that once I had the aircraft heading in a safe direction I was too low to eject. Despite the official attitude Jim Flemming was impressed and he convinced the Department of Air to allow him to make a 'Green' endorsement (a letter of commendation) in my flying logbook, even though he had recommended me for an Air Force Cross (AFC).

Despite the official attitude, the OC of the maintenance wing, Group Captain Cuming, wanted me to do a Test Pilots' Course. At that point, though, I was convinced I should resign. I felt indignant at the official reaction as I had done no harm, indeed, I had saved a five-million-dollar aircraft! My resignation was declined as it was claimed my services were still required. During a discussion with Group Captain Cuming, a highly experienced test pilot, I asked him why he thought the eagle had not disintegrated as most birds do at such high speed. His analysis was that, as the eagle approached the left engine intake, it disturbed the airflow. The airflow at the right engine intake was 740 kilometres per hour, which provided the pressure imbalance needed for a compressor stall. This is like a reciprocating engine back-firing. The effect was that the air racing into the right intake caused a back pressure at the left intake, virtually cushioning the bird as it was almost instantaneously accelerated to 400 knots (740 kilometres an hour) from the 9 or so knots it had been flapping along at. So the eagle entered the left intake at the same speed as the aircraft and suffered little damage on impacting on the stator blades. But it was too large to go between the compressor blades. Concussion of the impact, though, broke every bone in the poor eagle's body and this reminded me of the proverbial 'boneless chicken'.

THE RAAF AND VIETNAM

In 1966 a rumour was going around the base at Williamtown that the RAAF might be sending fighter pilots to Vietnam. However, no one knew in what role the pilots would be used. Based on this flimsy information, I made a written request to be considered. In those days the courtesy of an acknowledgement was always afforded one's request. My letter was acknowledged but it was vague and without commitment or admission that the rumour was even true. Doing all this operational training in the RAAF and experiencing what could only be considered low-level operations in Malaya, Borneo and Thailand, was frustrating. Not that I was desperate to go to war, but after so much training in the art of war, I figured that, logically, this training should be consummated with an actual combat experience.

In early 1967 I was put through a very boring Joint Warfare Course and given some limited training at controlling a Sabre onto a ground target while I was seated in a jeep. This was an artificial exercise but there was no one else in the RAAF who was experienced in the art of forward air controlling. My instructor wrote me up as having a natural aptitude for the task—not that he had had any experience in the field. With this in mind, I thought I would be in the box seat for a Vietnam posting if one came up. In October 1967 three other pilots were posted to forward air controller (FAC) duties in Vietnam. Since I had applied some time ago to go to Vietnam, I was a little surprised and quite upset about not being selected. In the meantime I suffered another incident flying the Mirage out of Darwin in the Northern Territory.

I was carrying out an engine air test just north of Darwin airport. During the high-power, high-incidence section of the test, the engine wound down. I was at 20 000 feet (6000 metres) at this point, which only gave me about three minutes of flight time before I would have to eject. As it was, I only had just enough time to turn a base leg at 5000 feet (1500 metres) and land. I felt I must have missed something during my checks and it troubled me that the engine would not start. The item I could correlate with my hurried checklist, was that the engine low-pressure fuel switch, which had come loose, fallen down inside the electrical consol and switched itself off. After this incident an AOG (aircraft operationally grounded) was raised that required the lengthening of the fastening screws on all Mirage low-pressure fuel switches. A previous modification had placed a guard over the switch but not replaced the old screws. This left only one thread holding the switch in place, allowing it to work loose.

Once again I was ridiculed for my decision to stay with the aircraft and this really made me livid. I had saved ten million dollars worth of airplanes and I was being criticised. The criticism coupled with the decision not to send me to Vietnam made me feel like I was wasting my time pursuing a RAAF career, as I seemed not to be able to do the right thing by them. I resigned my commission again but again it was declined on the grounds that I could not be spared from the job I was doing, which was as a senior pilot on 76 Squadron. All this made me despondent and I really did not have my heart in my job anymore.

In March 1968 the CO of 76 Squadron, Wing Commander Bill Horsman, called me into his office to tell me that I would be going to Vietnam as a FAC. Now I thought I could turn my failing career around. In early April Flight Lieutenant Roger Wilson, Flying Officer Macaulay 'Mac' Cottrell and myself boarded a Qantas Boeing 707 in Sydney bound for Singapore. On board were mostly US troops returning from R & R (rest and recreation) in Australia. We travelled first class and were treated like royalty. That night in Singapore we were given executive suites in the Lady Hill Hotel and the next morning were flown to Saigon, again first class, on a Pan Am Boeing 707. Chauffeur driven government limousines ferried us between airports and hotels. This was the way to go to war, I thought!

All three of us were rather excited about our new adventure and we were in high spirits during the trip. The contrast between the views we had seen out the aircraft window in Singapore and then Saigon was awesome. It was like landing in another world. Instead of the shiny clean aircraft, we now saw a mass of camouflaged military aircraft at Tan Son Nhut Airbase. I will never forget the strange smells and the humidity that hit me as I stepped from the air-conditioned Pan Am airliner. The lovely, smiling, sweet-smelling American hostess would be the last pleasant experience I would enjoy for some time. As we walked across the tarmac to the terminal we passed about 60 aluminium coffins stacked five high, awaiting loading onto 'The Big Silver Dust-Off', an endearing name given to USAF transports and airliners, or anything else for that matter, taking the GI home, dead or alive. This was a sobering experience.

Inside the terminal there were hundreds of GIs milling around, or just sitting on the floor, waiting processing. No one appeared to be in a hurry to go anywhere. Hell existed outside the doors, or so it seemed. Our Australian uniforms stood out like sore thumbs amongst the sea of US ones. Even so, no one showed much interest in us and it took us a couple of hours to find out where we had to go. It was as though the RAAF had got us to Vietnam and had done their job! I am sure I could have spent my whole tour just hanging around the airport and no one would have been any the wiser. Finally a kindly African-American military police officer drove us across to 7th Air Force HQ in his Jeep. His assessment, I guess, was that these guys are Air Force and should be at Air Force HQ. There we spent another hour or so convincing people who we were and what we were doing in Vietnam. Eventually we were shown to a USAF colonel's office. He was very pleasant and, after making a few phone calls, he established that we should report to the 19th Tactical Air Support Squadron (19 TASS) at Bien Hoa.

Bien Hoa was another major airbase situated 25 kilometres northeast of Saigon. Roger, Mac and I rode in a USAF Jeep through the crowded streets of Saigon and along Highway 1. The going was slow due to the number of people intermingling with bullock carts, all moving along the road in a haphazard fashion. The journey took an uncomfortable two hours. I felt very conspicuous and insecure in the open Jeep surrounded by masses of Vietnamese people through whom we had to negotiate a path. On arrival at the city of Bien Hoa we found the whole place in ruins and still smoldering from the Tet Offensive. Most of the buildings were crumbling or shot full of shell holes. At the air base there appeared to be no security, with civilians, bikes and bullock carts freely moving about. The Jeep driver kept honking his horn to literally ram his way through the crowds.

We were taken to the 19 TASS HQ, where the sign outside showed Snoopy sitting on top of his bullet-riddled kennel, shaking his fist in defiance. The greeting stated: '19 TASS—504 TASG, If you don't like the way I FAC, you can go FAC yourself'. Inside we were introduced to the amiable Lieutenant Colonel James T. Patrick, who was the CO. He welcomed us and outlined what would happen next and then handed us over to a sergeant, who showed us where everything was on the base and had us issued with guns and other kit. The kit consisted of: flack vest, 1 helmet and liner, 1 canteen and cover, 1 webb belt, 1 .38 holster, 1 flight suit, 4 tropical fatigues, 1 B-4 bag and 1 gas mask. This inventory was probably decided on during the Korean war, or earlier. I thought the gas mask must have been a carry-over from the Second World War. Later during my tour, when I found myself down wind of a gas operation, I regretted having left the gas mask in storage. I recall Jim Flemming telling me how he threw his gas mask over the side of the ship, when returning with troops to Australia after the Second World War. Many years later he received a demand for the equivalent of 10 shillings when an over zealous accountant found that a gas mask had not been cleared from Jim's inventory.

Finally we were taken to our temporary sleeping quarters. The beds were simple bunks with clean sheets and the building was a roof with flywire mesh walls. The sergeant showed us where we could eat and pointed out a bunker nearby to race into in the event of the siren sounding a VC attack. He said that mortars or rockets never hit in the same place twice, so our bunker was a good one. Someone had been killed in it only two days earlier when a 122 mm rocket went through the roof, spearing one of the occupants to the floor without exploding. This was no consolation for the poor GI who had a 122 mm hole punched through his chest. Under those circumstances I was surprised to get a good night's sleep; however, the effects of alcohol aided my shut-eye. Roger had *insisted* we consume a large volume of bourbon, scotch or rum to celebrate our arrival in a war zone.

At 6 a.m. all three of us had to be at the flight line to catch a C123 to Phan Rang Airbase. Phan Rang was a massive US Air Force facility located 280 kilometres northeast of Saigon on the east coast of South Vietnam. Here we were to do our indoctrination and FAC training on the Cessna O-1 Birddog. Phan Rang confirmed to us just how big the American machine was. To three young officers from the RAAF, the sheer magnitude of what we had seen since arriving in Vietnam was overwhelming. Each base we visited was many times larger than our entire air force. We arrived at Phan Rang on 13 April and had the now usual problem of identifying ourselves to others. Fortunately, this time we had US Orders instead of a meaningless piece of paper from the RAAF. We soon located the FAC school and were directed to our accommodation, which was the usual packing case and flywire decor. The bunks were double deckers with six pairs to a room. I could not find a vacant lower bunk so I threw my kit onto an upper bunk, closely followed by myself. We were not due to start indoctrination until the next day and I had decided to get some horizontal time. Despite the lack of air-conditioning or fans and the 100 per cent humidity, I had no trouble in sleeping through four hours of the continuous noise of jet aircraft taking off.

By late afternoon we were quite thirsty and, on Roger's insistence, we located the Officers' Club. The O-Club was situated on the highest hill at the base, with sweeping views, and about two kilometres from our hooch. A barbecue was almost permanently alight and we were soon engaged in conversation in the convivial atmosphere. I developed an immediate rapport with Major Amos Fox, who flew an F100 for one of the USAF squadrons on the base. He had a family back in the States and was getting 'short', that is, his tour was almost complete. He decided to leave as the club was starting to come alive. Amos said it was a trap to remain any longer as one got carried away drinking and then lost all sense of time. He had a dawn mission the next day and it was more important to him to be fresh for that rather than socialise deep into the night, and feel terrible next day. When I finally returned to our hooch Amos was asleep in the bunk underneath mine, such is the transient, gypsy way of life I'd begun.

The first day at FAC HQ we went through the indoctrination program. This consisted of lectures on how to get along with the local Vietnamese, films on all the known strains of venereal disease and many tips on how to stay alive 'in country', as it was known. It was pointed out that we had three enemies in South Vietnam and they were, in order of the most dangerous: friendly artillery, friendly helicopters, followed by the Viet Cong. It seems there were more deaths due to artillery fire from our own side, collisions with our own helicopters and other human errors than there were from being directly shot by the VC. Most of the lower level air traffic flew around at 1500 feet (450 metres), so you had to keep a vigilant look out for other aircraft. The transit height was selected as it was just above the small-arms envelope. So, one either flew above it or just above the trees, as you could not be heard approaching from any great distance. We were told we should consider everyone an enemy and we were told of many instances where the unwary were no more through being complacent about their personal security, wandering off into areas where they should not have been. Finally we were given the statistics of FAC survival. One in every ten of us would not survive, while one in every seven would be wounded. That was not what I was told before volunteering for this duty!

We started flying the Cessna O-1E on 17 April and my instructor was Captain Peterson. He was at the end of his combat tour and, as was usual, a pilot in this position completed his last few weeks in country as an instructor. The thinking was to take the pilot off combat while he was still alive rather than chance those last few weeks on operations. I was to learn of many pilots who continued on operations until the last couple of days of their tour, only to be killed when home was within their grasp. Captain Peterson talked me through all he knew about the aircraft as we sat in it and then we retired to the O-Club to see what else he could remember. Not much as I recall, given that the liquor soon flowed freely. The next day Peterson took me up in the O-1 to get a feel for the aircraft and then sent me on a solo flight after an hour of instruction. After that we retired to the O-Club to celebrate.

On staggering back to my bunk for an afternoon siesta I found two USAF officers going through Amos Fox's possessions. My first instinct was to aggressively demand to know what they thought they were doing. They told me that Amos had been killed this morning. I was stunned. They advised me that they were the Committee of Adjustment. When someone was killed a committee was formed to go through his personal possessions. This was to make sure military equipment was returned to store and that insensitive material was not released to the next-of-kin, and to prevent pilfering. Amos had been carrying out a GCA instrument approach for landing after completing his mission. On the turn to final he flew through the centre line, which took him too close to a 4500 foot mountain southwest of the airfield. His F100 clipped a ridge at 250 knots (460 kilometres per hour) and disintegrated. Although Amos was thrown clear and only lost his little finger, the fireball collapsed his lungs, which was the initial cause of death. Over the next couple of days I was to experience very eerie feelings climbing into bed over that empty bunk. It was soon taken over by a cigar-smoking slob, so I ventured off to find a more pleasant sleeping venue.

That evening Roger, Mac and myself were invited to the Australian O-Club, where No. 2 Squadron, Canberra bomber flight crews spent their spare time. The only person we knew there was Flying Officer Jock Bryant, an ex-Sabre fighter pilot who I had operated with in Malaysia and Thailand. Everyone else we thought treated us in a very cool, unwelcoming fashion. Roger and I thought that the cool reception was caused by the old fighter pilot/bomber pilot syndrome. Being new to the war, we did not appreciate the possibility that it may have been caused by battle fatigue or simply just a, 'leave me alone, I don't want to be here' attitude. Whatever it was, Mac decided to leave early while Roger and I kept the bar open until quite late. We could have gone to the USAF O-Club but I guess we were too lazy to move once settled in.

The walk from the RAAF O-Club back to our hooch was about two klicks by road, so we decided to take an illegal short cut past the 'Doughnut Dollies'—Red Cross Girls—compound. After passing the front door we found a six-foot-six, 250-pound (110-kilogram) lieutenant giving a young lady his full attention in the shadows around the side of the building. Both Roger and I were very much under the influence of alcohol and did not comprehend the importance this lieutenant placed on his activities. We tapped him on the shoulder and asked for directions to our hooch. At first he was very tolerant, under the circumstances, but Roger kept offering to help him with his task. I was the first to realise that this guy was going to get physical with us if we persisted in interrupting him and I dragged Roger away into the darkness. We spent some time navigating our way through the bushes and back onto the access track only to find ourselves back at the Red Cross Compound. On reflection, it would have been far more economical to have simply followed the road in the first place. A couple of days later the same lieutenant, whom we did not recognise in our sobriety, spotted us and engaged us in conversation. He was now in good humour and pointed out that we had in fact 'interrupted the coitus' and were within seconds of being beaten up.

The flying training continued in a relaxed fashion with only one or two one-hour sessions each day. I found Captain Peterson a bit of a cowboy in that, when he would demonstrate a manoeuvre, he would really throw the O-1 around and I shuddered at the sounds of the creaking noises coming from its overhead wing structures, protesting as he pulled on loads of 'g'. Perhaps he was merely displaying the 'dash' that George Turnnidge was always referring to! The O-1 carried eight 2.75 mm marker rockets, slung under the wings—two tiers of two under each wing. They were known as Willy Petes. There was no gun-sight for aiming them so you had to put a chinagraph pencil mark in the centre of the aircraft's front windscreen. You then had to manoeuvre the aircraft so this mark was on the target before firing. Depending on your head height, you would adjust the mark after firing a couple of rockets. After a few months I became very skilled and accurate and was just using my instinct rather than marking the windscreen. For training we used any abandoned huts, topographical features and even an old C47 crashed on the side of a mountain as targets.

Seventy-five klicks west of Phan Rang was the town of Dalat, which was 5000 feet (1500 metres) above sea level. The steep mountains facilitated a hydro scheme that provided power to the town and local area. Most of our training was conducted in the vicinity of this pipeline so we could double as protector of the facility. I never heard of the pipeline being attacked by the VC, probably because it was of as much benefit to them as it was to the general population and their supporters.

Another facet of our training was to learn how to fire the M16s out the window without doing damage to the aircraft. A number of overzealous pilots, getting carried away with the excitement of the battle, inadvertently shot their own spar away, causing the wing to come off with catastrophic results. On the first day both Peterson's and my M16s jammed. Mine fired two shots and Peterson's none. This did not fill me with confidence on the reliability of the M16. Some time later a US Army officer told me not to fill the magazine to maximum. By leaving it two bullets short it would prevent an inevitable jam at a crucial time.

On 21 April Captain Peterson left for the States, his tour complete. I was allocated to Captain Olsen, who was a little more fitting to my temperament and I enjoyed flying with him. He decided we should do some landing practice at Dalat. Being 5000 feet (1500 metres) above sea level it offered a cool change to the sweltering heat of the coastal plains. The serenity, the green vegetable fields and the fresh air impressed me and lifted my spirits. We were sitting on an ammo case having a cup of locally grown coffee, enjoying the peaceful, crisp morning air—we could have been a million miles from any war zone—when suddenly our reverie was shattered by incoming mortar rounds landing a couple of hundred metres or so from us. I jumped up and virtually ran in circles looking for a hiding place. Captain Olsen just sat on his box laughing. I felt very stupid and wondered why everyone except me was being so casual. At it transpired, the incoming was an almost daily event. The VC automatically fired four or five mortars every day, purely for nuisance value. Once the second mortar detonated the line of fire was revealed and, if it was not 'walking' in your direction, you just ignored it. Looking back, I am sure Captain Olsen planned this experience as a part of my training.

The next day I flew with Captain Weaver and did my final evaluation with Captain Johnson on 23 April. I picked an outcrop of rocks on a mountain slope to demonstrate my ability in delivering rockets. As I was pulling off the target I could hear gunfire and, without warning, Captain Johnson started firing his M16 from the back seat into the jungle. It was quite scary for me but Johnson just shouted out, 'Now, that's what ground fire sounds like!'

I was released for duty before Roger and Mac, probably because of my previous extensive light-aircraft experience. When I received my 'Certificate as Combat Cleared' I was on a C123 bound for Bien Hoa and I never saw Roger or Mac again. On arrival at Bien Hoa, having been up since the early hours of the morning to get there, I expected the rest of the day off. Captain Robertson had other ideas. He met me at the 19 TASS HQ and told me he was going to check me out on FAC-ing with fighters on a live target. I was assigned to Long Thanh North, also known as 'Bearcat' (YS 145982), some 30 klicks east of Saigon. Captain Robertson flew the O-1, with me in the back seat, from Bien Hoa to Bearcat.

Bearcat was basically a US Army aviation base where the lifestyle and conditions were halfway between a US Air Force base and an in-field US Army base. There was a wide variety of military aircraft there, including a Lockheed QT-2PC (Quiet Thruster) motorised glider. This glider could take off under its own power, climb to a few thousand feet and shut down its engine. It would then silently glide around doing surveillance work over US facilities, mainly at night, looking for enemy movement. When it descended to a low altitude, the engine would be started up again so it could be flown back to its original altitude to repeat the procedure.

This was my first experience in Vietnam of living on an Army base. I had thought that the USAF bases had been a little primitive in facilities but Bearcat was much worse. The accommodation was once again made from packing cases and flywire at best, or tents. The tents had loose timber-plank flooring, which was covered in mud or dust, depending on the weather at the time. The showers were very public, without walls, and the water pressure was gained by pumping water to a Korean War surplus fighter's fuel drop-tank on a stand. I soon learned the reason for the stampede to get to the shower first at nightfall. The water in the drop tanks was heated by the sun during the day. Later at night you would only get a cold shower. I did wonder why they bothered to put a roof over the shower, afterall, if you were showering with cold water, what did it matter if it rained! The toilets were semi-private, though, in that they had a low packing case wall around them and quaintly named the 'shitter'. The openness was very practical for ventilation! They were generally comprised of a bench with three or four holes in it, which were situated over individual 55-gallon (200-litre) drums without partitions between the occupants next to you. Some were six or eight 'holers', and you could not help but observe the toilet habits of the person seated across from you! Charming! It was also not unusual to have the drum removed from under you if the 'shitter detail' were doing their rounds, burning the contents of the drums. I don't know what you had to do to qualify for the 'shitter detail' but I imagine it had to be some horrendous misdemeanour. On combat operations I found that I was only getting to eat very small meals irregularly, so my need to visit these devices was, fortunately, only necessary every three or four days. The urinals consisted of an inverted funnel inserted into the top of a metal tube buried deep into the ground. These *l'appareil* were openly placed all over the base. As there were very few females around, publicly displaying yourself was not really a great issue. It reminded me of the dog and tree situation.

The day after I arrived Robertson demonstrated, with me in the backseat again, how to control three pairs of F100s in support of an Australian army unit situated in a dense rubber plantation to the south of Long Thanh. From the back seat I had little vision and not much idea of what was transpiring. There was a lot of confusing talk on several radio frequencies all at once, and I did not know who was talking to whom. Every couple of minutes an F100 would go roaring past us, ever so close, followed by the ear splintering explosion of a couple of 750-pound (340 kg) bombs. By the time we arrived back at Bearcat, I was totally baffled and suitably demoralised, not knowing how I was going to handle all this. I did not like being a FNG (fucking new guy), as all new soldiers in country were referred to, but this would not last long.

We had left very early in the morning for this mission and, as the aircraft had been refuelled and rearmed the night before, I did not consider it odd that the crew chief was not there. As it transpired, he had died during the night. He was a big placid type of person but I did not know his calm demeanour was produced by his round-the-clock consumption of whisky. Whether he did this to numb the sorrow of being parted from his family or the fear of war, or because he was an alcoholic, I never found out. He had gone to bed drunk, vomited while on his back and asphyxiated. Death was always just around the corner in this country.

The following morning I arrived at the flight line and was not looking forward to another day of demoralising confusion. While I was waiting for Captain Robertson to tell me what was going to happen today, Major Bill Walker arrived in an O-1 and told me I was to go with him, I was now assigned to his Brigade at Tan An (XS 525655), Long An Province. All sorts of things raced through my mind, such as Bearcat getting rid of me for incompetence. But I had not let on that I was confused and I had not yet been able to display inability, as I had not operated an aircraft in combat. In a warped way I was relieved to discover that the 3rd Brigade had just lost a FAC and an aircraft so I was to be the replacement. It gave me some respite in learning my new task. Being used to the Australian bureaucratic over-managment, I found it strange that someone should come along unannounced and simply lay down a new plan in such a short time span. As Major Walker was senior to me I grabbed my meagre kit, climbed into the back of his aircraft and went with him. So, my posting to Long Thanh lasted less than two days.

Major Bill Walker was a Texan, short in height but his high heel boots brought him up to average. He wore his side-arms low on the hip and his military hat was fashioned into a cowboy shape. My immediate impression wasn't great. I soon learnt, though, that his manner of dress was purely due to, 'that's what they do in Texas'; much the same thing as an English man wearing a coat and tie under a hot tropical sun. Bill's Texan accent over the radio was crisp, confident, clear and a pleasure to listen to. I developed a great deal of respect for this man.

We flew up to Bien Hoa where I was to pick up the replacement aircraft. The new O-1 was O-12438, but it only survived one week in Tan An before being shot down. I flew in formation on Bill Walker's wing to Tan An. At one stage during the flight he gradually descended lower and lower, making me believe he was going to give me some low-level formation practise but I did not relish the idea of transiting below 1500 feet (450 metres). In fact, he was going down to check out a target he had worked the day before. He made a number of passes over some destroyed bunkers at zero feet, which made me quite uneasy in this unfamiliar and hostile environment. As we climbed out Bill thought to mention over the radio, 'Y'awll need to get down low to see the size of their footprints!'

Unlike the usual pierced steel plank (PSP) runways made of Marston mats (named after the gentleman who invented them), Tan An had a bitumen surface—it was actually a fenced-off section of a narrow road. PSPs were strips of metal ten feet long and 15 inches wide and pierced by 87 holes. They were linked together to form the airstrip. Until the grass grew through the holes, the plates were quite loose and provided an unnerving rattle when the aircraft ran over them. Tan An's airstrip was three kilometres from the army base, which was sited to the west of the town of Tan An. At the airstrip was an Army field hospital, where it was convenient and expeditious to attend to badly wounded soldiers before shipping them out to larger surgical facilities. The road to town traversed open rice paddies and grass farmhouses were sparsely located along side of it. These houses became denser as we got closer to town and one of them was a brothel—it was not unusual to see a dozen or so GIs queued up outside waiting for their one US-dollar moment of pleasure. 'Not very romantic', I thought! To get to the army base you had to drive to the south-east side through the town. I never felt secure driving the distance between the airstrip and the base, it was always a relief to get back 'behind the wire'.

On arrival at our quarters I was assigned a bunk in the usual packing case and flywire structure with sandbags stacked a metre from the walls and up to two metres high to protect the occupants from shrapnel. The 3rd Brigade FACs had a hooch that was connected by a walkway to another hooch occupied by the 1st Brigade FACs. There was in all about eight or nine USAF pilots stationed at Tan An. Bill Walker progressively introduced them all to me as they came in off missions. The 3rd Brigade FACs were Captains Andy Anderson, Don Washburn and Ike Payne.

Andy and Don were on rest in the hooch whilst Ike was out flying. Andy was a quiet, unassuming type with a pleasant demeanour. He had previously flown the F86D with the USAF and was a graduate of USAF flying training program. Don was a little more on the reserved side and did not communicate so readily but was still a first-class officer and a likeable person. He had been flying previously as co-pilot on the B52 and was a West Point graduate. We were all engaged in meaningless conversation when Ike Payne came in and I was a little startled by what followed. A voice behind me entering the hooch said, 'Hi ya Honkies!' Which was responded to by a chorus of, 'Hi, ya black bastard!' and, 'Hey Ma, the nigger's home!' I was used to white Americans being a little less abrasive with black Americans but this was different. I soon realised it was all in good humour but not to be practised generally between coloured and white people who did not know one another so well. Ike Payne was also a graduate of West Point. He was a little overweight, friendly and jovial. I took an instant liking to him and he was to be my instructor pilot (IP), taking me through my combat training.

In addition to flying, we all had secondary duties to perform. Initially, I was assigned as navigation officer and tea club officer. Wow! My operational call-sign was to be Tamale 35, as mentioned earlier. 'Tamale' was the call-sign of all 9th Division FACs; 'Tamale 3-' was the 3rd Brigade whilst 'Tamale 1-' was the 1st Brigade and 'Tamale 2-' the 2nd Brigade. 'Tamale 30' was the 3rd Brigade TACP, 'Tamale 31' was the senior pilot or ALO, and the call-signs 'Tamale 32' onwards depended on seniority. The FACs were forever changing call-signs as new pilots joined the brigade. As the foreigner, I managed to maintain Tamale 35, which suited me fine as it was one less thing to think about. As it was, Tamale 35 became synonymous with my name and even today, 38 years later, Americans still refer to me as Tamale 35.

The following morning was 27 April 1968, and at 5 a.m. Ike Payne took me up in the back seat for my first real 'on the job' training flight. As we were climbing out to the southeast of Tan An there was a series of deafening explosions. We were only two minutes into the flight. Ike immediately threw the O-1 into a 90-degree banked turn, slamming the throttle fully forward while screaming, 'Fucking Gooks!' I had no idea what was going on but the look on Ike's whitening face made me think I should be terrified. And terrified I was. I thought to myself, 'How can it all be over so soon?' Ike was on the radio demanding answers in no uncertain terms. I soon learned that this was one of those 'friendly' enemies. ARVN artillery. US artillery bases would broadcast on a special frequency their intentions to fire, but the ARVN would just open up when they felt like it. And we had flown into an ARVN artillery barrage. The thinking, in cases such as this is when you have not yet been hit, is to circle hard at that position and find out where the artillery is coming from. After finding out you could then estimate which direction to fly to get clear of it. Now better informed, Ike picked a heading and the deafening explosions receded behind us. Many times during my tour I was to experience this uninvited 'friendly' fire and it always brought my heart into my throat.

After escaping the friendly artillery, Ike explained what he was doing and why as we flew all over our area of operation (AO). He briefed me on the location of various US facilities, hot enemy areas and anything else he thought I should know to keep me alive. On arrival back at Tan An, Bill Walker asked how I felt about putting in an air strike by myself. This surprised me a bit as I had not even put in an air strike under supervision and only witnessed someone else put in three. But what the hell, 'Yes, I can do that!' I said with confidence.

Bill arranged an air strike for me at the junction of the Vam Co Tay and Vam Co Dong rivers—grid coordinates XS 607 733—which was a known VC hive. It was with some apprehension that I took off and headed east towards a new experience. Two F100s checked in and gave me their ordnance load as 750 high-drag bombs and RPs (rocket projectiles). By emulating what I had experienced with Captain Robertson a couple of days before, everything worked out well and I was pleased at my advancement in the world of 'FACing'. The air strike destroyed four bunkers and six structures that were hidden in the nipa-palm.

I was met by Bill Walker when I arrived back at Tan An and he told me that to shorten a pilot's training time, he would always pick a target safely away from friendlies where, if there was a stuff-up, no damage would be done. He would monitor the radios from the TACP radio room to see how the strike progressed. Unknown to me the whole TACP were listening to my efforts and basically grading me. As I had sounded confident and proficient, Bill Walker was happy to release me and said, 'Good work, Cooper—you're going to make a fucking fine FAC.' I must say that this experience cleared up all the doubts I had in my own ability and enhanced my confidence.

The next day I couldn't wait to get out and try my newly developed skills. The pre-planned strikes were in the area of the 'Testicles', so-named as the Vam Co Tay River did a double reversal on itself, producing what looked like a pair of testicles when viewed from the air. After reconnoitering the target area I was ready for my fighters.

'Tamale 35, this is Magpie 61.'

Oh Shit! I thought. This was going to be different. Magpie 61 was a Canberra aircraft and these bombed in level flight from 5000 feet (1500 metres). They ran in from about eight kilometres out and were hard to keep in sight due to their distance from the target. They carried a good bomb load but were slow in positioning themselves for the run in. If there was any urgency to get bombs on target, we would always send the Canberra high to hold while we used up any dive-bombing fighters we had available. I am sure the Canberra pilots used to get annoyed at being asked to come in last but they had plenty of loiter time which the fighters did not have. Further, the accuracy of their level bombing from high altitude was not as good as dive-bombing, particularly in strong crosswinds when the smoke marker would drift downwind. Having said this, I must add that the Canberra bombers were operated by very professional crews but the aircraft were more suited to saturation bombing. Due to their shortcomings in this type of combat, I did not like using them in contacts close to friendly troops. On this particular occasion, there were no friendly troops nearby so it was good to practise putting down ordnance at a leisurely pace.

Over the last two days of April I consolidated my learning by controlling two flights of two F100s each day using 750-pound bombs and napalm. This accounted for the destruction of eleven bunkers and five structures.

4

THE MAY OFFENSIVE

ALTHOUGH NORTH VIETNAM AND the Viet Cong started planning the Tet Offensive in July 1967, it was not until 29 January 1968 that it actually began the simultaneous attacks on US facilities throughout the whole of South Vietnam. This included the attack on the US Embassy in Saigon. The casualties from the offensive were as follows:
- 3895 officers and men of the US Army, US Air Force, US Navy and Marine Corps
- 214 officers and men of the Republic of Korea Forces, Vietnam; the Australian Task Force, Vietnam; the New Zealand Army Force, Vietnam; and the Royal Thai Military Assistance Group, Vietnam
- 4954 officers and men of the Republic of Vietnam Armed Forces (South Vietnam)
- 14 300 civilian men, women and children of South Vietnam
- 58 373 officers and men of the Vietnam People's Army (North Vietnam) and the South Vietnam People's Liberation Armed Forces (Viet Cong).

The Tet Offensive officially ended on 31 March 1968 with the defeat of the communist forces. However, history now indicates

THE MAY OFFENSIVE

that the communist considered the 'defeat' was more a strategic withdrawal and in early May contact with the enemy started to become frequent again. I was new in country and not aware of any changes in the combat situation. However, as the intensity increased I wondered how I was going to survive a tour flying combat every day around the clock. It was a trial of physical endurance. As it turned out I flew anything up to 16 hours a day in any 24-hour period. The month of May witnessed some of the largest and most costly battles of the Vietnam War. The May Offensive, along with the Tet Offensive, made 1968 the bloodiest year of the war.

I started just before dawn on 1 May, flying Captain Andy Anderson, whose call-sign was Tamale 32, up to the maintenance base at Bien Hoa where his aircraft was being serviced. To get to Bien Hoa from Tan An, we tracked straight up Highway 4, over the Ben Luc bridge to the southwest corner of Saigon. We then followed the Kinh Doi Canal along the southern edge of Saigon, up the east side of the city to Long Binh, then direct to Bien Hoa. We would follow the reverse of this route to return to Tan An, which kept us clear of airline traffic out of Tan Son Nhut and directly over artillery support bases. It was always safest to fly directly over the artillery bases, as there was a cone overhead where we were out of the line of any outbound ordnance and reasonably safe. Although US artillery would constantly broadcast on the artillery net where they were firing from and to, we would sometimes miss broadcasts due to cockpit workload and other distractions and find ourselves in the middle of an artillery barrage. After dropping off Andy I headed back to Tan An back along the same route.

On approaching the southeast tip of Saigon, Tamale control advised me to put in two pre-planned air strikes near the Ben Luc bridge and that the fighters were on the way. I barely had enough time to reconnoiter the target when the first set of fighters checked in on my frequency. They were a pair of F100s carrying 750-pound high-drag bombs. We opened up some bunkers without any return ground fire. The second fighters were also F100s armed with 500-pound slicks and napalm, which destroyed some more bunkers and four sampans I had located among the reeds in a small stream. I was beginning to feel quite familiar and at ease with my role as a FAC.

While I was refueling the aircraft back at Tan An the crew chief, a Herculean-built African-American with a big 'chip on his shoulder' called Jake Johnson, an Airman First Class, or E-3, came over to tell me I was to get airborne straightaway to support troops-in-contact between Can Giuoc and Rach Kien, about ten klicks south of Saigon. Jake would always start a conversation with, 'Shit man!' Most of the time, that is all he had to say. This time he had strung a few more words together. 'Shit man, ya gotta git airborne—da troops in contact, man!'

'Whereabouts, Jake?'

'Shit man, I dun know!'

To the regulated Australian mind of the RAAF it might seem strange that I should jump into an aircraft and take off after such a brief statement by an unauthorised airman without rank. But it has to be remembered that time was a critical factor in war and all I needed was an indication of which way I should head after take off because everything that followed was handled on the radio and orchestrated by me. In the RAAF I would have to be correctly authorised and briefed by an appropriate officer before contemplating jumping in the aircraft! The saving of lives was paramount and I really liked this no-nonsense approach of the Americans in their efforts to get the job done.

For this target the fighters were two Cessna A37s armed with 500-pound bombs, napalm and cluster bomb units (CBU). This was my first experience with the A37. Known as the 'Dragonfly', it was a small twin-engine training aircraft with side-by-side seating. Being a slower and lighter aircraft than the heavy fighters, it could stay in close to the target and, in its early introduction to Vietnam, was still being flown by very experienced evaluation pilots. I found them extremely accurate and a pleasure to work with. The enemy quickly withdrew and disappeared with only an

THE MAY OFFENSIVE

initial show of light small-arms fire. It struck me that the enemy did not want to stay and fight. We would later learn they had a more important task in mind.

In addition to air strikes, I was flying a further four to six hours each day on VR (visual reconnaissance) missions, convoy escorts and administration flights. On 2 May I was doing one VR mission along Highway 4 but I did not detect anything unusual. At the morning intelligence briefing the briefing officer said that there were massive enemy build-ups taking place in all sectors. When the Army stumbled upon any enemy, the enemy units soon dispersed and could not be found again, indicating that they were massing for a major confrontation.

The runway at Tan An was a section of a narrow bitumen road culminating in a bridge at the western end. The bridge was sealed to traffic to keep the locals out, but to no avail. We often saw Vietnamese riding bikes along the runway and I wondered how many locals I must have just missed in the dark during night landings. One day as I lined up on final approach to land the control tower fired red flares into the air, then I spotted a bike rider in the middle of the runway. He was taking no notice of the flares so I made a very low pass at him, head on, but he still kept riding on at a leisurely pace. I made my landing approach, passing my wing a few centimetres above his head, but he still peddled on. After landing I did a 180-degree turn to return to my revetment and taxied straight at the rider, who did not flinch or give way, showing every indication that he considered I was not there! I suppose he had travelled that section of the road all his life and he was not going to let a little war change his habits.

On 3 May I had two pre-planned air strikes scheduled for targets five klicks southeast of Saigon. The country was very flat and muddy with little vegetation except along the banks of streams, so these were the areas I searched for signs of the enemy. Being just on sun up there was a ground fog and the air was beautifully smooth. The Birddog literally slid through the air and it was pleasant flying along, rolling the aircraft from side to side as I followed the streams in the cool morning air just above the dispersing ground fog. This target was generated by Intelligence and I must admit, I was very sceptical when I first arrived there that anything would be found. But I located five structures and 17 bunkers among the vegetation, which we destroyed with five pairs of F100s carrying 750-pound bombs, napalm and 20 mm cannon. We received no ground fire, so if the enemy were there they did not want to make contact.

When one of the sets of fighters checked in he transmitted, 'Tamale 35, this is Devil 21 . . .' I could not understand what he was saying after Devil 21, even though I had him repeat his transmission three times. It sounded as though he were speaking another language. In frustration he eventually said, '*Buenos dias*. G'day you Aussie bastard—Joe Turner here!' He was throwing in a bit of Mexican Spanish to go with my Mexican cuisine call-sign. Colonel Joe Turner had been on F86 Sabres with the RAAF as an exchange officer from the USAF. He had completed his tour with the RAAF some time before I left for Vietnam. 'Devil' was the call-sign of the 614th Tactical Fighter Squadron of the 35th Tactical Fighter Wing (TFW) and Joe Turner was the squadron's commander.

Like the previous day, during the pre-planned strike south of Saigon on 4 May, we saw no sign of the enemy but we did destroy four hidden structures with two F100s. This air strike was followed by three hours of VR but I could find no further evidence of the enemy. It was already dark by the time I landed at Tan An. When I arrived there was a USO show under way with Bob Hope and lots of leggy dancing girls. One of the entertainers was Tony Sheridan, who had performed with the Beatles in their early days. They had accompanied him when they were called The Beat Brothers. In fact, I had his record called 'The Beatles First', recorded live in Hamburg, Germany, and it featured the Beatles as musical accompaniment, not doing any vocal work. Tony Sheridan was English and he had married an Australian girl, making Australia his home. When he discovered I was Australian, he spent most of the show talking with me, distracting the visual attention I was rightfully giving to the girls. I naturally asked him the question that everyone else did—after rising to fame working with the Beatles, why Vietnam? He pointed out that he was receiving $400 for a ten-minute performance—doing four performances a day, seven days of the week—all expenses paid, and the gig was tax-free. That ended that line of questioning! Tony still performs rock music today and lives in Germany.

Apart from the infrequent USO shows, which were hard to synchronise with my flying schedule, there was a base cinema that I never attended as well because my flying operations never dovetailed with the session times. There was also the Armed Forces Radio and Armed Forces TV. The Armed Forces TV ran limited programs but everyone would try to catch the weather report. It was read by a delightful looking young blonde girl called Bobbie, who tried to put on the 'dumb blonde' act but I am sure she was intelligent. Bobbie was officially in Vietnam as a secretary to the US Agency for International Development and lived on Nguyen Hue Street, Cholon, for three years. Her stunning looks meant she was in high demand as a morale booster and, in the course of her duties, she escaped gunfire, slept in bunkers, flew in helicopters and rode on buffaloes and all manner of livestock. She was even catapulted off the USS *Enterprise* in a fighter. All this was to show that she cared about us and to spread cheer among the troops in remote places. Bobbie even gave up her R & R to visit troops in the field. So many women served with distinction in Vietnam but received no recognition. I am sure no-one could remember what she said about the weather, but all wished that they had a job with Armed Forces TV. Every time Bobbie mentioned rain, and this was the wet season I should add, someone off camera would throw a bucket of water on her. She took to wearing a skimpy bikini if she knew she was going to mention rain. If she knew she was not going to report rain during her presentation, she would wear a neat civilian uniform. While trying to present the weather in a dignified manner, someone unseen by the viewers would be reaching up her skirt, causing her to prance around and giggle. Bobbie visited Tan An while I was based there but I was out flying when she came and missed the vision.

As history now tells us, the renewed Offensive started on 5 May. On this date I was scrambled at 2 a.m. to Binh Phuoc, Long An Province. The army base there was under ground attack from a large enemy force. This was my first operation at night and I had no idea what to expect. When I arrived over Camp Robert Rethune I could see the mortar rounds impacting on the army base and the flashes from the 2nd Battalion, 4th Artillery returning fire. Mixed with this were the telltale orange flashes of the Soviet-made 122 mm rocket exhausts as they left the darkness outside the army base. One advantage of night flying was that I could see all the rounds being fired by their flashes, which enabled me to position myself out of the line of fire. Small arms, though, were a different matter as it was the muzzle flash you did not see that got you. Due to the movement of the aircraft the enemy firing at you had to allow a lead on his target to get the right trafectory, so you did not see the muzzle flash if the gun was not pointed at you. If you saw the muzzle flashes the rounds would be going behind you. The exception to this general principle was when you were diving on a target. If you saw muzzle flashes straight ahead, then a hit was imminent. Although, often when I was diving on a target the tracers appeared to be coming straight at me but at the last moment they would curve to one side, over or under the aircraft. This gave me a false sense of security, as the tracers did not always curve away! Artillery, mortar and rocket fire required a different consideration. Even though you could see where they were coming from, you still had to note where they would explode to work out which bearing to avoid. Further, the distance the rounds were landing from the point of launch would give you an idea how high the 'max ord' was—that is, the highest altitude of the ordnance. Sometimes, depending on what operation you were involved in, it was best to fly underneath the rounds instead of above them. Then you had to be vigilant and watch for any new player firing from a different direction. I was not fully aware of any of this on the morning of 5 May. I had not received any instruction in this facet of war, so I was about to receive my first, big, self-taught introduction into working with artillery on-the-job, and at night!

On my way to Binh Phuoc, Tamale Control gave me the call-sign and frequency of my contact at the target. By the time I arrived overhead I already had a good idea where most of the enemy elements were.

'Big Gun, this is Tamale 35 on 88.5 fox mike.'

'Tamale 35, go.'

'Understand you need some TACAIR on your perimeter,' said I, hoping artillery would not be requested.

'Negative, Tamale, we cannot wait for TACAIR. Can you adjust our artillery for us? We have been putting out suppressive fire but cannot get the range. It looks as though the point of fire is two to three klicks to the north.'

Blast! I thought. 'OK, Big Gun. I have noted you receiving fire from the south and east as well. We will not go for flares or markers, as we want to surprise the Gooks. Standby for coordinates to adjust your arty.'

It was a pitch-black night and I had to really work hard to prevent becoming disorientated and losing control of the aircraft. I had my map in my lap, a flashlight in one hand and a Chinagraph pencil in the other. At the same time I had the control column clenched between my knees and was pushing the rudders with my feet in order to control the aircraft. All this was taking place while my sight was focused simultaneously on the map, the instruments, Binh Phuoc and the three targets I had observed, and I had to write the coordinates on the side windscreen.

I was now frantically trying to remember the phrases I had heard over the radio from someone else directing artillery.

'OK, Big Gun, I have your coords—ready to copy?'

'Tamale, go.'

'XS 568 615—XS 550 549—XS 538 606, read back.'

'Roger, Tamale. I have, XS 568 615—XS 550 549—XS 538 606.'

'OK, Big Gun, give me five rounds HT mixed with two rounds VT fusing on each coord simultaneously.'

'Standby, Tamale.'

A few seconds later, 'Tamale, this is Big Gun, firing for effect.'

'Big Gun, go!'

Although I had no experience in this type of work, it all seemed so natural for me. All the time this was going on the enemy fire continued. I noted that most of it was from the north, south and east so I held my position directly over the army base with the heavy missiles going through underneath me. I was flying at 500 feet (150 metres) with one window open. The sounds were loud and frightening. In a matter of seconds the barrage was over but there was still some enemy fire coming from the north.

'Big Gun, this is Tamale. Let's have one round at a time on XS 568 615 and I will move it around a bit—fire when ready.'

Almost immediately there was a loud 'kaa-woompf' below me.

'Big Gun, add 100,' which meant increasing the point of impact by 100 metres.

'Roger, Tamale.' Another 'kaa-woompf' followed.

'Big Gun, right 50', meaning, move the impact 50 metres to the right. There was no explosion.

'Big Gun, that was a dud. Give me another one please, same correction.'

'Roger, Tamale.'

The 'kaa-woompf' was followed by an enormous secondary explosion and a fireworks display that lasted for a couple of minutes. The jubilation over the radio showed a considerable relief of tension on the ground with many 'shit hots' and 'real fines' broadcasted. I was amazed at how easy the whole thing worked out. To the accolades I merely, and embarrassingly, said, 'Big Gun, you pushed the button. I only told you how hard,' in the most confident manner I could muster.

After the exchange of pleasantries I checked in with Tamale Control on Victor to get clearance to leave the target. Having received the clearance I climbed to 5000 feet (1500 metres) and flew around for 30 minutes on minimum power so that the enemy could not hear me. They would think I had left the target. If the enemy had been listening on VHF, and they probably were, my clearance home would have further confirmed that I had left. After 30 minutes there was no renewed attack so it was safe to assume the task had been completed. I slowly descended back to Tan An as the sun was coming up. The air was beautifully smooth and as I slid earthwards, with the sudden release of tension, I had trouble keeping awake. On landing it took me only ten minutes to hit the bunk. I was exhausted.

On 11 March 1998, just on 30 years later, a friend of mine received the following email from Ron Van Dyck who had been on the ground at Binh Phuoc on the morning of 5 May 1968:

Date: Wed, 11 Mar 1998 07:46: 14 0800 (PST)
X-Sender: pao_armyla@earthlink.net
To: btate@nor.com.au
From: Ron Van Dyck <pao_armyla@earthlink.net>
Subject: Cooper

Dear Brian,
I served with Btry B, 2d Bn, 4th Artillery (105mm towed) from Nov 1967 until Aug 68, then at HHB 2/4 until I DEROSed at the end of Oct 68.

I don't remember ever meeting Garry face-to-face. Might have, though on a night when the whiskey was flowing.

I was stationed at Binh Phuoc, Camp Robert Rethune, in Long An Province. Binh Phuoc was a firebase/patrol base for our battery and the 5th Bn, 60th Infantry (Mechanized). My assignments were: FO, FDO, XO, CDR of the battery, in that order.

During the dry season of 68, the firebase was attacked several times by 82mm mortars. We did several analyses, set an observer on top of our FDC, and the Infantry had a guy up in a tower, sniper bait, all designed to locate the mortar and its crew. About 2.8 kilometres up the straight road that ran from the northern boundary of the base was where we reckoned Charlie was operating, firing 5-6 rounds then disappearing into a tunnel. We silenced the mortar one night with 7 charges of white phosphorous rounds fired in that direction with times set on the fuses bracketing the bad guys. The howitzer crew had pre-positioned three rounds between their trails, with the tube set at the guessed elevation. When our observer saw the flash, the XO gave the command to fire all three rounds and get into their bunkers. Apparently we got them, because the mortar never fired again.

About a month later, there was a big enemy operation in the area, and an aerial observer spotted some VC scurrying and disappearing into an opening. I think I was at Tan An with 2 tubes from my battery and the Battery Commander was at Binh Phuoc with 4 tubes. (We used to do that to support ground operations.) At Binh Phuoc 2 x 8-inch SPs (M-110) had rolled in from a GS unit to help support the operation as well.

Cooper was in the air, and in our forward FDC we listened to the mission. (My 2 tubes were out of range.)

Cooper was directing fire from three artillery elements at three different locations. One of the 8-inch rounds failed to explode, and he asked for another round, same place. He got it, and both exploded, then all the ammo in a cache there exploded. He was getting secondary explosions for about five minutes. That was the cache that the mortar VC were using.

I was very impressed with Cooper's ability to direct fire from multiple elements, and keep track of the GT lines and co-ordinate the 105, 100, and 8-inch fire.

That's probably the most memorable event for me of my Vietnam tour. That happened in April/May 1968, and I think that blowing up that cache probably saved the lives of many soldiers at Binh Phuoc.

Regards,
Ron Van Dyck.

Having come off duty at dawn on 5 May, I was not scheduled to fly again that day. My next flight was to be early evening on 6 May but at about 2 p.m., while I was gainfully filling in my time taking a nap, I was scrambled. My destination was Long Phu Tay, four klicks northeast of Can Giuoc, twelve klicks south of Saigon. Here Bravo company, 3rd Battalion 39th Infantry (3rd/39th), had

THE MAY OFFENSIVE

been ambushed by a company-size enemy force equipped with AK47s and B-40 rockets. As it transpired, this enemy force was a company of NVA soldiers; that is, uniformed soldiers from North Vietnam and not Viet Cong. This was unusual and it alerted us that something big was about to happen. History now reveals that the VC forces had been almost wiped out during the Tet Offensive earlier in the year. Their ranks were being replaced by NVA soldiers from North Vietnam.

First Lieutenant William L. Forrester had been leading his Bravo Company across open mud-lands on a reconnaissance-in-force mission when they were ambushed by the enemy who were well-entrenched in bunkers in a line of palm trees running east–west across Forrester's line of approach. The mud-lands were abandoned rice paddies with not a stick of vegetation for cover—in such areas the local people actually buried their deceased family above ground, placing the coffin on a mound above the high-water mark and setting a concrete and stone grave with headstone over it. Some of Bravo Company had managed to gain cover behind three such graves. The rest simply squirmed there way into the mud where they'd dropped for cover.

Tamale Control had ordered three sets of fighters for me. On arrival over the contact area I had a few minutes to take stock of the situation. I could see the tracks left through the mud by Bravo Company and judging by the drag marks, the poor devils must have been up to their knees in the muck. These tracks ended abruptly in a small circle where the soldiers had gone to ground, literally, and buried themselves. I could also see about three soldiers behind each gravestone. My circling around had brought some relief to them as the NVA stopped firing for fear of giving away their position; however, Forrester dared not move his men, as they would again become prime targets. The ground situation was such that it was obvious to me where the enemy were and Forrester's description confirmed this. It was ideal for me as I did not have to get the friendlies to mark their position with smoke and I had a clear picture of exactly where the enemy was concealed. To confirm my estimates I flew at 50 feet (15 metres) along the treeline the enemy were in. Forrester confirmed I had the right treeline but I could see no sign of the enemy and they resisted the temptation to fire at my low-flying aircraft. As I pulled off the treeline, the UHF radio burst into life.

'Hello, Tamale 35, this is Dice 22.'

'Dice 22, this is Tamale 35, go ahead,' I replied as I climbed back up to 500 feet (150 metres).

'Dice 22 is a flight of three F100s carrying 500 and 250 slicks and napalm. We are presently about 10 klicks north of your position.' As I was turning left though north I picked up the tell-tale black smoke of the JP4 burning fighters.

'OK, Dice, I have you visual. Keep following the river under your nose south and I will tell you when to turn. Are you ready to copy your briefing?'

'Go ahead, Tamale.'

'Dice 22, your target is XS 866 755. It is an enemy force entrenched in an east–west orientated treeline. We have friendlies 200 metres to the south in the open. Attack direction will be from west to east with left turns. You can expect automatic weapons fire up to 50 cal. Best bailout area will be over Can Giuoc, four klicks southwest. We will use your slicks first and then burn them.'

Dice One, the leader, read back the required points then said, 'Dice 22, line astern, go.'

'Two.'

'Three.' The other two aircraft had acknowledged then moved into a trail position on Dice One.

'OK, Dice, start a left turn now and I will be in your 9 o'clock at the centre of your turn.'

'We have you, Tamale, and ready to go to work.'

As soon as the fighters started turning over the target, the enemy decided there was no point trying to remain concealed. They opened up with all the firepower they had. I clearly heard the automatic-weapons' fire and I was now their prime target. Their thinking was, take me out and the fighters could not rain fire and brimstone upon them.

THE MAY OFFENSIVE

'FAC in for the mark,' I transmitted as I rolled in from the west. I had chosen to make my pass directly along the treeline, which would make me a difficult target due to the trees offering some screening. To attack at right angles to the treeline would have been foolhardy as I would be in full line of site to the whole enemy force.

I'd put my smoke in the centre of the tree line, which was about 400 metres long. I then cleared the fighters to bomb anywhere along the tree line. As I flashed over the trees at 100 feet (30 metres) I could hear the continuous rattle of the enemy guns and my body tensed, expecting to feel red hot metal going through me. No hits—they were notorious for not allowing enough lead, thank God! I turned right over the friendlies, looking over my shoulder as Dice One called, 'Lead, in hot.'

'Clear hot, Dice One,' I replied.

Dice 22 put down their slicks so accurately that by the time they finished with the napalm, the target was obliterated and quiet. Just as I cleared Dice 22 from the target area, a flight of three Vietnam Air Force A1 'Skyraiders' checked in.

'Tarmly Tee Fyv, dis Wylat Sic Sic.'

When translated, 'Tamale 35, this is Violet 66.'

'Hello, Violet 66, this is Tamale 35. Are you ready to copy briefing?'

I had trouble understanding what he said in response as I had difficulty understanding their accented English. But I took it that he did not want a briefing. This was frustrating as I needed to spread the next ordnance around to catch any escaping enemy. 'OK, Violet, standby for my mark', I replied, pushing on.

I rolled in and put a marker about 100 metres behind the treeline. As I pulled off the target I gave Violet the clearance to bomb, telling him that I needed to spread the ordnance, only to find Violet One had already rolled in, very steeply, and was releasing a pair of 500 slicks right behind me. These bombs hit my smoke exactly. Before I could say anything Violet Two was in a dive and put his bombs in the same hole as Violet One. I saw little advantage in trying to direct them so I circled over the friendlies, keeping alert in case I needed to give the Vietnamese pilots a 'Stop, Stop, Stop' if they started to drift from the target area. They were deadly accurate, unfortunately, as all their ordnance, including their napalm and 20 mm cannons, went into the same hole! With this, Violet One transmitted a leaving message. 'Ve go now, tank yu Tarmly.' All I could think to say was, 'Good bombing, Violet. Nice working with you.'

Many years later I again found myself in a pickle due to a misunderstanding in language, in rather more humorous circumstances. In the 1980s I was flying a Boeing 747 aircraft for a company in Europe, operating all over the world on charter work. This meant I frequently found myself in different countries. On one occasion I was with a crew in Seattle, Washington State and head office in Luxembourg advised us to go to Santiago, Chile, to take a flight to New York via Mexico City. My first officer was running short on out-of-base-time (you could only be away for so long) and had to go back to Luxembourg. A replacement first officer was located in Miami and the hotel receptionist there left the verbal message that he should go to Santiago. However, the receptionist wrote down that the location was San Diego (California). The flight engineer and myself duly arrived in Santiago; the first officer did not show up at our hotel. We thought he had taken accommodation at the airport to get more rest but when we arrived at operations in the morning we were advised that the flight would be delayed 24 hours while they got the first officer down from San Diego.

Some of the VNAF pilots had 5000 or 6000 hours on A1s and over six years on ground-attack operations, so they were very capable and extremely accurate, but the language barrier was a real problem at times. However, as long as I put in an accurate mark I was never concerned about working with them in close proximity to friendly troops. The VNAF used the single-seat A1H Skyriders, which were different to the standard A1Es. The A1Es carried a crew of two and a very heavy bomb load that meant using full power for takeoff. At full power the massive Pratt & Whitney R-3350 engine produced an enormous amount of

THE MAY OFFENSIVE

torque, which was quite within the capabilities of an American to handle. For the Vietnamese, who were generally smaller in stature, it was often beyond their physical capabilities. So the VNAF used A1Hs and, at the lighter weights, needed only METO (maximum except for takeoff) power that enabled them to handle the torque. Even then the engineers had to build up the rudder pedals so that the pilots could reach full throw on the rudder.

As soon as I finished with the A1s, two A37s checked in.

'Tamale 35, this is Rap 21.'

'Rap 21, go.'

'Rap Lead, we have been overhead and did not want to disturb you when you were having so much fun. (A slight chuckle could be detected in his voice.) We are two A37s carrying Mk82 slicks with mini-guns and we have the target in sight.'

'Roger, Rap. From the treeline we were attacking I would like to move back to the north spreading the ordnance around to catch any retreating Gooks. Put your slicks down in an area 100 metres square, using the A1s' napalm smoke as the southern boundary.'

'Got it, Tamale.'

What a breath of fresh air, I thought.

'Rap One, you're clear hot.' We then worked over any retreat areas with the bombs and mini-guns.

After I cleared Rap 21 from the area I called Lieutenant Forrester on FM. 'Hunter, this is Tamale.'

'Go ahead, Tamale.'

'That's all we have for the time, do you require further TACAIR?' I could see Bravo Company were on the move again, widely spread and heading north towards the target, so I flew low over the bombed area. The devastation was incredible with hardly a tree left standing.

'Tamale, this is Hunter, no more TACAIR required for now. If you can hang about until we get into the treeline I will give you a sitrep [situation report].' As they moved toward the treeline I circled around the general area at low level looking for other footprints or evidence of movement.

'Tamale, this is Hunter.'

'Go ahead, Hunter.'

'Christ, Tamale, there's nothing left down here. It's all blown to hell. That was a fine job and saved our ass. I will send you an AK47.'

'My pleasure, Hunter. See you soon.'

Instead of departing the area, I climbed up to 5000 feet (1500 metres) so that any enemy could not hear me circling and would assume I had left. I hung around for another 30 minutes until my fuel was getting low and landed back at Tan An 3 hours and 50 minutes after I had taken off. The results of this contact were 15 bunkers destroyed, 5 structures destroyed, 22 enemy killed by bomb action (KBA) for 3 wounded in action (WIA) US soldiers.

5

THE 'Y' BRIDGE IN SAIGON

GOOD TO HIS WORD, Lieutenant Forrester sent the AK47 to Tan An for me. It was delivered in person by the Commanding Officer of the 3rd/39th, Lieutenant Colonel Anthony P. De Luca who sent me an email recently thanking me for my air support during the May Offensive. He wrote:

From: Tdeluca53@aol.com [mailto:Tdeluca53@aol.com]
Sent: Tuesday, January 11, 2005 8.55 AM
To: tamale35@bigpond.net.au
Subject: 1968

At the battle at the 'Y' bridge in May 1968, I didn't know you by name—but I did know that crazy FAC who was helping our troops by flying low and slow to mark the targets for the fighter-bombers. That action helped prevent a lot of casualties. I was delighted to sign an endorsement for the Silver Star for you for that action. Warmest regards and thanks for saving our collective a__es!,

Tony De Luca.

Tony De Luca was a unique officer. He commanded a battalion in combat in Vietnam with the disability of having sight in only one eye. In July 1968 his light observation helicopter (LOH) was shot down. His pilot was killed with a 50 calibre wound to the head and his artillery officer suffered a broken back. Tony spent some harrowing time evading the enemy and returning to his ground unit after getting his artillery officer onto a Medivac chopper. When the LOH was recovered it had almost 50 small arms hits on it. He became the fifth Colonel shot down in two months. The order of the day was keep away from Colonels! The AK47 he delivered for me was a rather gruesome object as it had not been cleaned up and was still covered in mud, human flesh and black hair. This enemy weapon is now on display at the Vietnam Veterans' Museum at San Remo in Victoria.

Another survivor from my actions with 3rd/39th on 6 May was Sergeant Howard E. Querry. Unfortunately he was to die in combat under heroic circumstances only four days later, on 10 May during another action in which I was involved. His widow, Pauline Laurent, was seven months pregnant at the time of his death and recently wrote a heart-rending book, *Grief Denied*. This was all about her struggle as a war widow and brings to light the anguish suffered by the next-of-kin of slain soldiers. The soldiers suffering in combat is short compared to a lifetime of anguish suffered by bereaved relatives.

On 7 May I was scheduled to put in a pre-planned air strike four kilometres south of the Ben Luc Bridge, just east of the Van Co Dong River. The intelligence section, based on information from many sources about enemy movements, generates these pre-planned strikes. Being pre-planned, there was no rush so it was a pleasant change. All I had to do was have a target ready when the fighters arrived at the rendezvous. I always tried to allow myself a minimum of 15 minutes to reconnoitre a target area before the fighters arrived. With no urgency on this strike—because troops were not in contact and it was preplanned based on sketchy information—I was able to take my time to check over the aircraft, taxi out and take off. After starting I did a casual engine run up,

THE 'Y' BRIDGE IN SAIGON

took the runway and eased the throttle forward to full power. As the O-1 gathered speed along the section of public road that was our runway I played with the rudder, aileron and elevator controls to smoothly lift the aircraft off the ground. Once airborne I climbed to 4000 feet (1200 metres) and headed toward my assigned area.

At this height the air was cool and smooth and I enjoyed the scenic flight. On arrival over the target I maintained my altitude and surveyed the area through my binoculars, hoping to catch some enemy in the open. An O-1 at that height on minimum power cannot be heard from the ground. Not seeing any indication of movement on the ground I descended to 100 feet (30 metres) above the trees to tempt any enemy there to fire upon me. There was no reaction. I then selected an area of nipa-palm that I considered would be the most likely place for an enemy base camp. By slowing the aircraft down to just above the stall with 30° flap, I was able to weave around and get a reasonable view down between the vegetation. Sure enough, I spotted some structures, bunkers and discoloured foliation, indicating recent activity in an otherwise relatively remote location. I increased to full power, pulling in the flaps, commencing to climb out of it, noting the coordinates of the location as XS 632 719.

On my way up to 1500 feet (450 metres) the fighters checked in. They were two F100s carrying 500-pound slicks and napalm and we were soon at work. It was basically a training mission as we knew there would not be enemy action and we were able to run a neat left-hand pattern and take our time practising pinpoint accuracy. After the fighters had expended their ordnance, I went down to 50 feet (15 metres) and circled the target. The napalm had burned off the vegetation and I was able to give the fighters a BDA of two structures and four bunkers destroyed.

The fighters left the scene and I commenced my climb out, heading back to Tan An. Suddenly Tamale Control came up on the UHF to advise me that there were troops-in-contact only three kilometres south of Saigon. The contacts were getting closer to the capital but the significance of this did not occur to me at the time.

The coordinates placed the contact at Xom Chong on the Ragh Xom River, which was a small stream running south off the main Kinh Doi Canal along the southern edge of Saigon city. Tamale Control had already ordered the fighters for me, and the first set checked in on my frequency before I arrived at the target area. It was fortunate that the Cessna A37s had a bit of loiter time unlike the heavy-metal F100s, which had to be put to work quickly. I had the A37s hold at 10 000 feet (3000 metres] over the south-west tip of the city, while I flew at full throttle to the target. The coordinates placed the target right under the departure path of the airliners and big jets taking off from Tan Son Nhut Airport, which was Saigon's main airport. So already a restriction was being placed on our attack direction. With this knowledge, I was able to commence a part of the briefing even before I reached the target. As the big jets were taking off towards the southwest, they were then turning left and flying around the south side of Saigon before taking up an easterly heading out over the South China Sea. I decided to have the fighters attack from the west into the east with right turns. This enabled the fighter pilots to look toward Tan Son Nhut on their downwind leg and have a good clear view of their flight path and any possible intersecting traffic. The A37s were carrying 500-pound slick bombs. Slicks had to be released from a higher altitude in a steep dive, which meant they would be diving through the airliners' level. The napalm was not a problem as this ordnance is released from low level at a shallow angle, which meant working below the airliners' flight path.

I contacted the Grunts, US Army soldiers supposedly named from the sound the men make when lifting up their rucksacks and slopping around in mud. They are also known as 'Bluelegs', in contrast to artillerymen, known as 'Redlegs'. I found that they were not really pinned down but needed the TACAIR to get rid of a company of enemy blocking their progress. It was far more prudent to use TACAIR than risk friendly lives with a ground engagement. The enemy were entrenched in heavy nipa-palm on the east bank of the river. As I approached the area I had the Grunts put out red smoke and they gave me a bearing and distance to the

THE 'Y' BRIDGE IN SAIGON

enemy. As the fighters were high, and I had not yet reached the target, the enemy did not know that they were about to be bombed. The timing was perfect. I placed my mark into the nipapalm and the ground commander confirmed I had the target just as the first A37 was rolling in for his attack from 10 000 feet (3000 metres). It all happened like clockwork and the enemy would not have known what hit them as the perfectly placed pair of 500-pound bombs hit my smoke marker exactly. I had the A37s expend their ordnance on the target in a pattern that provided good coverage. The ground commander confirmed 18 enemy KBA by body count—that total did not include those who would have been blown to pieces.

As the A37s left the target area, 'Magpie', an Australian Canberra bomber, checked in. This was not my first experience with the Canberra so I knew what to expect. The Canberra generally level bombed from 5000 feet (1500 metres) with a conventional bombsight. Due to their method of delivery and antiquated equipment, it took some time to get their ordnance on target. However, they were quite accurate and carried a good bomb load of 500 and 1000 pound (225 and 500 kg) bombs. As the Canberra needed a long run-in for their bombing strikes it meant that an airliner could enter the target's airspace while Magpie was running in. Another problem was that Magpie's bombs would need be released from above the airliners. I was forced to have two dry runs by the Magpie bomber due to crowded intersections with heavy jets departing Tan Son Nhut. The ground commander did not need any further TACAIR, but as I had Magpie available, I decided to soften up areas further north where the army unit was intending to reconnoitre. This proved to be a worthwhile decision as the ground commander later reported a further 15 enemy dead in the area of Magpie's bombing.

I felt rather good about the achievements of the day and was back at Tan An before dark. I was even back in time to attend the mess hall for dinner for the first time since I'd arrived. Although I needed the sustenance of a cooked meal, the standard was such that I preferred my staple diet of cold baked beans, or creamed corn and biscuits taken from supplies back in the hooch. I had become accustomed to this boring fare. Because of this meagre diet, and constantly sweating in the aircraft, my weight reduced from 85 to 62 kilograms within two months of arriving in Vietnam.

There had been some excitement back at Tan An while I was away during the day. A soldier went berserk and began shooting his weapon indiscriminately. The result was that three GIs were killed and a fourth wounded before the murderer was restrained. It is thought that his actions were drug related but there were many cases of soldiers just snapping under the stress of combat. Thankfully this was the only incident of such bizarre and tragic behaviour I encountered during my tour.

The next morning I had a pre-planned air strike south of the Ben Luc bridge in an area we called 'The Bowling Green'. It was a region of dense rice farming, 15 kilometres long, spreading from Highway 4 and bound by the Vam Co Dong River, over which the Ben Luc bridge spanned to the north. It was bound in the south by the Vam Co Tay River, which flowed past Tan An. It was one of the few areas where the local population felt safe in their communal numbers to carry on farming. In most other areas rice farming had fallen into neglect. The Bowling Green owed its name to its rich green symmetry that resembled a bowling green from the air. We were to have many enemy contacts in the Bowling Green because it turned out to be a staging camp for the NVA, a place where they could rest up, be well-fed by the locals and hoard their weapons. All this was done under the noses of various American agencies that ironically provided aid to the local population. The NVA would enter this area from the Ho Chi Minh Trail via the Parrot's Beak on the Cambodian border, and then by way of the Kihn Bo Bo Canal that was a straight, man-made canal 60 kilometres long, traversing thousands of acres of abandoned rice-growing country. We now know that this was the route the enemy followed to attack Saigon and to supply their war effort in the southern provinces of South Vietnam. On this air

THE 'Y' BRIDGE IN SAIGON

strike I was using two A37s with wall-to-wall 500-pound slicks. The A37 is pretty to watch due to their accuracy, but on this mission all we destroyed were four bunkers with no enemy response.

That evening I was scheduled for another pre-planned strike within three kilometres of Saigon. Once again I had A37s armed with 500-pound slicks and napalm. They also carried their own flares and that saved me calling in a flare ship or using artillery delivered flares. It was interesting to watch. One would put out illumination while the other aircraft attacked. As I could not get any enemy to fire on me, we attacked the intelligence coordinates but I could not give the fighters a BDA because I could not see the results in the dark. With everything quiet, I left Saigon and was back in bed at Tan An by two o'clock the next morning.

That afternoon, 9 May, I carried out another pre-planned strike, just to the south of Saigon, with three F100s carrying 750-pound high-drag bombs. We opened up eight bunkers with surprisingly no enemy reaction. After putting in the air strike I was directed by Tamale Control to fly support to a ground convoy proceeding north up Highway 4. This convoy was on the way to Long Bihn, which took a couple of hours, so I had to refuel at Bien Hoa before heading back to Tan An.

At six o'clock on the morning of 10 May, I was scrambled to assist Alpha and Charlie Companies (A/5/60 and C/5/60), 5th Battalion, 60th Infantry (5th/60th) who had been ambushed five kilometres south of Saigon on Highway 5A just north of the village of Xom Tan Liem. The commander of the 5th/60th, Lieutenant Colonel Eric F. Antila of Santa Fe, New Mexico, had tasked Alpha and Charlie Companies to carry out a recon in force (RIF) south of the Kinh Doi Canal along Highway 5A. Alpha Company led to investigate the reported presence of a Viet Cong company in the area, with Charlie Company following along Highway 5A. Alpha Company came into contact with the enemy, which turned out to be an NVA Battalion, not a VC company! Veterans of Charlie Company told me at a reunion in Denver on 16 June 2001 that they all felt very uneasy about this patrol and the

Charlie Company commander, Captain Edmund B. Scarborough of Belle Haven, Virginia, even questioned Colonel Antila about his judgement in sending them out unsupported. Colonel Antila was determined in his resolve and Scarborough, being a dedicated 28-year-old soldier, followed the orders of his commander. When the patrol was one kilometre north of the Xom Tan Liem, they were caught in a classic ambush by a well-organised and heavily armed enemy force.

Alpha Company became pinned down and Captain Scarborough decided to charge the hamlet along the road. Unfortunately, what he didn't allow for was their inability to leave the road. The road was raised with deep water-filled channels on either side. The lead track with Captain Scarborough on board was immobilised by an RPG and, simultaneously, so was the rear Armoured Personnel Carrier (APC). The commander had a few seconds to reflect on his mistake, realising that there was not a thing he could do to recover the situation. All the soldiers dismounted to take cover but they had enemy fire coming at them from both sides as well as from up the road. Alpha Company had taken cover in some of the houses on the west side of the road, but Charlie Company was in an exposed position. While Captain Scarborough was taking cover behind his destroyed APC another RPG hit him from behind. Scarborough was killed, as were three of his men. They were William G. Behan of Philadelphia, Pennsylvania, Richard J. Flores of Hanford, California, and Randolph R. Wilkins of Camden, New Jersey. Captain Scarborough was posthumously to receive the Distinguished Service Cross, his nation's second highest award, for his bravery and conduct during his seven months of combat.

It was about this time that I arrived on the scene. 'Bandido Charlie, this is Tamale 35 on Fox Mike 25.2.'

Silence!

'Bandido Charlie, this is Tamale 35 on Fox Mike 25.2', I called again. Just when I was about to switch to my VHF and ask Tamale Control if I had the right frequency, a concerned voice came up in reply.

THE 'Y' BRIDGE IN SAIGON

'Tamale 35, this is Bandido Charlie.' It was the voice of Lieutenant Charlie Taylor. Two F100s fighters-bombers checked in on my UHF radio and were carrying 750-pound high-drag bombs, which pleased me, as the delivery of these was ideal for close contact. From my observations, I already knew that the fighters would have to attack from the north, parallel with the highway, and I briefed them accordingly and told them to expect ground fire. By this time I was switching between my FM, UHF and VHF radios, talking with five different agencies simultaneously. It was a demanding task to keep track of the conversations while flying the O-1 at the same time.

I had to keep the conversation flowing with Bandido as I was obviously speaking to someone who was not used to talking with pilots in a TACAIR situation.

'Roger, Bandido, you need some TACAIR?'

'Ah, Tamale, this is not my normal job. All my superiors are dead and I have had to take command.'

'OK, Bandido. Where are you getting fire from?'

'All quarters Tamale, but the worst is from the east side of the road about 100 metres south.'

Just then another voice came up on 25.2 Fox Mike and it was the beleaguered Alpha Company. 'Tamale, this is Bandido Alpha One. I can confirm that, and we are directly across the road from the Gooks. We are in some hooches and are pinned down.'

'OK, Alpha. Throw a red smoke onto the road in front of you.'

'Roger, Tamale.'

'Alpha, I have your red. Where is the nest in relation to that?'

'Fifty metres east, Tamale.'

Hell, I thought to myself, this means we will be bombing within only 75 metres of the nearest friendly troops. As I was positioning for my marker pass another voice came up on Fox Mike. This time it was Colonel Antila in the command chopper, who had only now been able to get to this contact, which was quite understandable when you consider the demands placed upon a battalion commander with all his units involved in conflicts in different areas at the same time.

'Tamale, this is Iron Guard. I will hold out to the west until you lay down your TACAIR and, be advised, you are being fired on by 12.6 mm anti-aircraft guns from one klick to the west.'

'Thanks, see if you can get some artillery on them from the west and north to keep the max ord off the line of attack of my fighters, who will be running in from the north.'

'Will do, Tamale. I will have Nickle Nuthin' handle it.'

Who the hell is Nickle Nuthin'?, I thought. I found out later Nickle Nuthin' was Army talk for 'Five Zero' and Colonel Antila was referring to his forward observer (FO), who was on the Raven OH23 with him. The FO's task was to control any artillery requirements and assist his commander. Antila's FO on that day was Lieutenant Tommy Franks, who later gained fame as a four-star general. He was to serve as the Commander in Chief of the United States Central Command and led the American Coalition Forces to victory in Afghanistan and Iraq.

Antila's arrival had added another thing to my list of considerations, so my envelope of attack was getting even more restricted. There was not enough time to brief the fighters of this extra danger of incoming artillery so it was my task to make sure they did not fly into this new no-fly zone. I decided that an accurate way of getting a marker onto this target was to go in at zero feet and drop a smoke-grenade out of the window by hand. This action was fraught with danger but I could not afford to put down an inaccurate mark. It had to go right into the lap of the enemy.

'Tamale 35, in for the mark.'

As usual, all the radios went quiet and it is almost as though everyone involved was holding their breath to see if I would be shot out of the sky. At 500 feet (150 metres) I rolled on 120 degrees of bank, sliding the nose below the horizon to place the O-1 into a steep dive. Once I had the dive-angle set, I quickly pulled the safety pin on a smoke grenade by taking my left hand off the throttle, holding the grenade and the control column with it and pulling the pin with my right hand. Then still holding the grenade and the safety lever, I placed my left hand back on the throttle, transferring my right hand back to the control column. I could

THE 'Y' BRIDGE IN SAIGON

not afford to fumble at this stage! It's a bit like fielding at first slip in a cricket match, but with dire consequences for a dropped catch. That done, and with my target in sight, I started weaving the O-1 from side to side by kicking the rudder around in order to try and make myself a difficult target. Along each side of the road were tall coconut palm trees and I levelled off almost with my wheels between them. Although I could hear small arms fire, I had a silly false feeling of security, because I was out of the line of fire of the heavier 12.6 mm anti-aircraft guns. As I had been running in from the north along the road, I flew over the pinned down Charlie Company, and at tree-top height I could see the upturned faces looking at what they prayed would be their salvation. I said to myself, 'Don't stuff this one up, Coops, as everyone down there is relying on you.'

Just as I was approaching the t-intersection in the village I could see enemy troops to the eastern side of the road. I threw the Birddog into a 90-degree left bank, took my hand holding the smoke grenade off the throttle, hung my arm out the open side window, and let the grenade go. I slammed the throttle full forward and whipped the O-1 into an opposite steep right bank, away from the enemy occupying the east side of the road. Looking over my right shoulder I observed the smoke billowing among the enemy troops, who were starting to run away from the road. Most of these troops were not the usual black 'pyjama-ed' Viet Cong, but NVA—uniformed North Vietnamese soldiers who were professionally trained.

The fighter aircraft had positioned themselves in a left pattern in preparation to attack from the north. With all the ground fire present, I would have preferred to give them random attack headings where they could roll in from any direction they wanted, thus keeping the enemy guessing where they were coming from. However, this was a highly restrictive target and a complicated ground situation so the attack direction had to be from the north along the road. I realised this would give the enemy an advantage because when they heard the approaching aircraft all they had to do was fire into the north. Consequently the fighters would have to fly through a curtain of fire, greatly increasing their risk of being hit. I believe there were more aircraft shot down in Vietnam by flying into a wall of lead than there were by the traditional tracking method of delivering fire at a moving target. I quickly gave clearance for Bobcat 21 Lead to attack.

'Bobcat Lead, clear hot with 750 pairs, the Gooks are scattering to the east, put your bombs 50 metres east of the base of my smoke.'

'Roger Tamale, I have your mark—in hot.'

By this time I had completed a 180-degree turn and was back up to 200 feet (60 metres). Bobcat Lead was head on to me, only separated horizontally by the width of the road. As Bobcat released his high-drag bombs from 100 feet (30 metres) I was positioned to turn in behind him and watch the bombs sail down among the fleeing troops. Many of them would have been vaporised instantly. I then cleared Bobcat 2 to attack.

'Bobcat 2, clear hot. Put your bombs down 20 metres left of Lead's.' By doing this correction, I was following the direction of the retreating NVA troops.

'Roger, Tamale. Two in hot.'

Once again I was able to turn in behind Bobcat 2 and watch his bombs fall among the enemy. This time, however, I climbed to 400 feet (120 metres) as the explosion from Lead's bombs were deafening and the O-1 really bucked in the concussion. Just after Bobcat Two's bombs detonated, Bandido Alpha called up on Fox Mike, 'Hey, Tamale, those bombs are collapsing the roof on us!'

'Alpha, do you want me to back off a bit?'

In true Rowan and Martin 'Laugh In'-style (the 1960s variety show that started Goldie Hawn's career), he said, 'Negative, Tamale. SOCK IT TO 'EM BABY!' The relief of the tables being turned on the enemy must have triggered this levity.

In quick succession I had Bobcat release their remaining ordnance around the general area to get maximum coverage. While I had Bobcat on target, two other sets of fighters checked in so I had them stacked up to the south. By briefing the new

THE 'Y' BRIDGE IN SAIGON

fighters in between controlling Bobcat, I had the next fighters rolling in after Bobcat pulled off the target. This workload kept my mind off the ground fire and I hardly noticed it.

The second set of fighters were A37s carrying napalm and cluster bomb units (CBUs), which are ideal against troops in the open. The CBU consists of bomblets measuring 2.75 inches (seven centimetres) in diameter and 10 inches (25 centimetres) long. The body of each bomblet is made from soft metal and contains hundreds of 10 mm steel ball bearings. The centre of the bomblet is filled with high explosives. An impact fuse on the nose of the bomblet ignites it and when it explodes, the ball bearings fly out in all directions at high velocity. As there are several hundred bomblets in each of the two 10 feet × 3 feet (3 × 1 metres) containers attached to the aircraft, they spread over a distance of up to 500 metres and literally chew up anything on the surface. To deliver the CBU the fighter flies low and level over the target. When the pilot actuates his release switch, the covers over the front and rear of the canister fall away allowing the slipstream to force the bomblets out of their cylinders. Simple but effective!

Now that we were about 150 metres from the friendly troops I started using my Willy Pete rockets to mark the targets, as the smoke hand grenade was a rather hazardous method of delivery. The third pair of fighters were F100s armed with 500-pound bombs, napalm and CBU, which was another very applicable load of ordnance for this target. Now that we were starting to get the enemy out into the open, the operations sections were regularly arming the fighters with CBU and napalm. I was moving each attack to a different location, trying to get maximum coverage of the ordnance. I was firing a rocket for each pass of the fighters and was getting low on Willy Pete rockets by the time the third set of fighters had left the target.

I contacted Tamale Control to see what the availability was of further fighters and was told that there would be none for at least two hours, due to a sudden increase in enemy activity. For the next 30 minutes I flew at 100 feet (30 metres) around the contact area and attracted no ground fire and was advised by Bandido

Alpha and Charlie that all opposing fire had ceased. With this information, I climbed up to 5000 feet (1500 metres) and circled for another 30 minutes, all the time staying in contact with the ground forces. It became obvious that any surviving enemy troops had retreated, otherwise they would have attacked the vulnerable US troops again when they thought I had gone. The army count was 20 VC/NVA dead, which did not include those vaporised by the USAF bombing and napalm. There were also three VC prisoners taken and Charlie Company felt quite deprived of some sort of revenge, when Lieutenant Colonel Antilla and Tommy Franks landed in a chopper and took them away for interrogation.

Among the survivors of this battle was Sergeant Alan Kisling, an 18-year-old from the back woods of Oregon. In April 1999, out of the blue, I was to receive the following email from him:

From: "ALAN" <AKISLING@THEGRID.NET>
To: "Brando Madrigal" <letup@email.msn.com>,
"Garry Cooper" <GarryCooper@compuserve.com>
Date: Tue, 30 Mar 1999 12:03:43 -0800

Hello!
I am Alan Kisling, aka Shorty. I was a Recon Sgt of 'C' Co, 5th Bn (mech), 60th Inf Regt., 9th Inf Div from 10/67 to 10/68. From what I understand, our paths have crossed. If I have events and time line correct, you, Mr Cooper, are a major factor in my still being alive and well. Without your air support we would have lost our entire company. I owe you big time!
THANK-YOU,
Shorty.

In 2001, at a reunion in Denver, Alan Kisling told me of how they collected the dead and wounded from their units before regrouping further north toward Saigon. The details brought tears to my eyes. I was subsequently made an honorary member of the 5th/60th. Alan, his lovely Mexican wife Sylvia, my wife Jean and

I have developed an enduring friendship and we have spent many enjoyable hours together in the United States.

I landed back at Tan An at 10 a.m. after four hours in the air and immediately crashed on my bunk, only to be woken at 2 p.m. to do a four-hour visual reconnaissance (VR) along the Bo Bo canal, up to the 'Parrot's Beak' of Cambodia. I flew at my usual high altitude on minimum power, but I saw no activity. At the time it was well known that the enemy had already settled into the areas south of Saigon but it was necessary to keep a watch on the Kihn Bo Bo to stem any resupply efforts made by the NVA. It was starting to get dark when I arrived back at Tan An so I missed mealtime again, thank goodness! After devouring a can of cream-corn and biscuits, I headed back to my bunk. Unfortunately rest was denied me. At 9 p.m. I was woken again with troops-in-contact. The coordinates were Saigon itself!

Sergeant Cover, our radio operator mechanic driver (ROMAD), told me what he knew as I pulled on a fresh flight-suit, laced up my combat boots and grabbed my weapons: one colt .38 six-shot revolver and an AR15 5.62 mm carbine with six spare magazines—the revolver was really only useful to use on yourself if you were captured and the carbine was to shoot as many of the enemy as you could before you were. I was in the jeep within ten minutes of being woken. It was then that Sergeant Cover said, 'No hurry, Sir—Captain Anderson is not back yet with your aircraft. He should be landing in about ten minutes.' 'Great!' I thought. I could have had another 15 minutes in the sack!

Sitting in the jeep at the revetment, I first noted the artillery stop firing, closely followed by the sound of the little O-1 touching down. No sooner had it landed than the artillery opened up again, making me realise how critical the timing was for our landings and take offs.

Captain Andy Anderson taxied in and even before he had shut down, the ground crew were on the wing refuelling the aircraft. Andy jumped out, shook hands and I climbed aboard and strapped in. As I was strapping in Andy briefed me on what he had experienced during the contact near the 'Y' Bridge in South

Saigon, which he described as bedlam. He also said that he had to skirt around a developing thunderstorm midway between Saigon and Tan An on the way back. It was with some apprehension that I continued with my preparations. The likeable Andy slapped me on the back as I hit the starter and shouted above the engine noise, 'Take it easy, Buddy!'

As I was doing my magneto check I called Tan An artillery and asked them to hold firing while I took off (the magnetos run the spark plugs and always have to be checked that they are working correctly). This had to be well-timed, as there was a window of only two minutes given for the aircraft to take off and clear the firing line before they would start firing again. There was no warning given when they restarted, so you had to make sure you made it out in that two-minute window. There was a light breeze of 10 knots (18 kilometres per hour) from the west and as I rolled down the narrow road, the runway lights began to flash by with increasing speed. Halfway through the take-off run it started to rain lightly and I noticed for the first time lightning flashes over my shoulder to the northeast along my intended flight path. As I pulled the O-1 off the ground and pointed its nose into the blackness of the night I noticed the shadow of my aircraft being thrown on the ground to my left by the high intensity flashes of the lightning coming from my rear starboard quarter. Standard procedure on taking off from Tan An was to fly straight ahead at a steep angle to at least 1000 feet (300 metres) before turning. An early turn would put the aircraft low over the heavily foliated banks of the Vam Co Tay River, which was infested with Viet Cong. Low flying aircraft would be sure to attract ground fire.

Just as I started my turn right I experienced severe turbulence and, despite having full power on, my airspeed reduced alarmingly and my altitude decreased to 500 feet. In the middle of all this the control tower radioed that they were experiencing 65 knots (120 kilometres per hour) of wind speeds from the east. There was my problem, wind shear, giving me virtually no airspeed due to the sudden change of wind being brought on by the approaching thunderstorm. I fought to maintain control of my aircraft and

THE 'Y' BRIDGE IN SAIGON

continued the turn to get into the 65-knot headwind. On approaching my climb out heading, and having only just recovered from the negative wind gradient, what I saw ahead almost stopped my heart. In the lightning flashes I could see a wall of green water rising from the ground to a high altitude like some monster from the deep. It would have been foolhardy to try and manoeuvre out of its way, as the heavy torrential rain instantly consumed the Cessna. I could see nothing. With no sight of the ground, with the heavy rain and wind battering the little aircraft and with the lightning flashing around me, it was like being inside an incandescent drum with someone beating it with an iron bar. The turbulence was the worst I had ever experienced, so much so that my feet were thrown off the rudder pedals by the negative 'g' forces. I really expected the aircraft to break up or for me to lose control. All I could do was to lock onto my artificial horizon and try not to change the power too much in an effort to chase the airspeed, as it was going anywhere between zero and 200 knots. I flew through this terrifying environment for perhaps five minutes when all of a sudden I popped out into a clear, star-filled sky. As I had been too busy to make any radio calls, I called Tamale Control to receive details of my strike and exact coordinates.

Tamale Control said, 'Glad you're still with us 35, that must have been a rough departure.' With my heart still pounding, all I could think to say as casually as possible was, 'Roger, that!'

'35, this is Control, your coords are XS 843 883 and you have troops in heavy contact. I have ordered three sets of fighters for you and you should contact Raven 66 on Fox Mike 26.6 over the target—good luck!'

Writing the details on my windscreen I replied, 'Thanks, Control.' Plotting these coordinates on my map confirmed that the contact was the Y Bridge on the southern canal of Saigon City. As I approached Saigon I could see a mass of tracers crisscrossing in all directions, the flashes of artillery impacting the ground, the anti-collision beacons of numerous helicopters and the illumination of the flamethrowers from the Monitor craft on the canal. With the passage of the recent storm, the air was heavy and caused all the smoke to hug the surface, making objects on the ground hard to discern from the air.

'Hello, Raven 66, this is Tamale 35.'

An apprehensive voice came back shouting to be heard above the explosions. 'Tamale, this is 66. We are pinned down with Gooks all around us, as close as 50 metres. Can you help us?'

With a positive air I said, 'Certainly can, 66. We will be bombing close in so I want you to hug the ground until I have finished. Can you pop smoke at your position?'

'Not much point in that, Tamale. Each time we pop smoke the gooks put down the same colour.'

I decided to use a colour code we had recently discussed at briefings. It meant mentioning all the colours available but the friendlies would only put out the second colour mentioned and the enemy a different one. There would be a delay in the enemy smoke while they searched for the correct colour.

'OK, 66. If red is white and purple is green, I want you to pop yellow.'

'Roger that, Tamale,' the ground commander came back without hesitation, so he was aware of the code. The thing in the code that would confuse the VC was that my first transmission was just an alerter, the friendly smoke was not to be popped before a command to do so. Immediately I had mentioned 'pop yellow', a ring of yellow smoke appeared on the ground, so it was a good bet the friendlies were at the centre of this circle. Sure enough, when I commanded, 'Raven 66, pop yellow', white smoke appeared at the centre of the yellow circle.

While all the preceding was going on I was also briefing three sets of fighters stacked up overhead, as well as trying to ascertain where the artillery was coming from and listening out on the Tan Son Nhut frequency as the airliners were still exiting Saigon right over the contact area. As the Army could not afford to have the artillery lifted, we had to work with it passing through our attack space and hope that we would not be hit. The airliners were flying out in a blaze of light with landing, navigation and anti-collision lights on. We could see them clearly as they approached, and the fighters were able to dive past them taking their own avoidance action with only verbal inputs from myself. As the airliner passengers were nearly all military personnel, I suppose it would not have concerned them as much as it would have had they been fare-paying passengers.

The artillery impacting into our area of operations was coupled with numerous flares that lit up the whole area. The flares were going off without warning or control. They were raining down among the fighters and the O-1, blinding us and subjecting us to disoriention at times. The whole scene was chaotic. Due to the amount of traffic and smoke, I had to fly below 500 feet to keep the target in sight.

Raven 66 was not the only unit requiring tactical air support but they were the most in need at this moment, as assessed by central control. There were helicopter gun-ships close by supporting other ground units. I had the fighters set up to run in from the west parallel with the Kihn Doi Canal, breaking right for a right pattern over relatively unoccupied air space. Using the colour smoke reference from the friendlies, and the yellow smoke kindly provided by the enemy, I dived in for my mark sending off a white phosphorous marker rocket 50 metres from the beleaguered US Army unit. The mark was good and I was able to direct the fighter to 'hit my smoke'. As I banked left, all of a sudden I had to roll on 90 degrees of bank and pull back hard on the control column to avoid a Huey that filled my windscreen. Just as I rolled back right to keep the attacking fighter in sight and fly back over the target, the helicopter exploded less than 100 metres from me. It all seemed to happen in slow motion. First I noticed an unusual glow coming from the helicopter that grew into a fireball, consuming the craft. The only part not engulfed by fire was the rotor blade, which stuck out of the ball of flame, rotating upwards and flying off on its own into the darkness. The fireball descended vertically impacting the ground and burnt fiercely in the centre of what looked like a large playing field, right at the southern exit to the Y Bridge. Artillery or an RPG may have struck the Huey, but whatever it was, those inside would not have known what hit them.

The attacking fighter anxiously transmitted, 'Was that you, Tamale?'

Silly question—if that was me I would not have been in a position to answer!

'Negative—hit my smoke.' I guess his adrenaline had been really pumping, but for some reason, this incident did not really shake me. I was probably too preoccupied with the task at hand. Many years after the war, I reflected on my response during such situations. It occurred to me that it was all relative, and I was always more composed when someone else was being dealt a worse deal than myself. An example of this relativity to which I am referring was when I dropped out of the 12.6 mm anti-aircraft fire and felt relatively safe with only small-arms fire to contend with! However, it did remind me to keep weaving to deny the enemy the ability to be able to track me as a target.

The fighter released two cans of napalm right on my target and I was able to direct the rest of the ordnance from these two F100s using the flames from this first attack. The Huns were followed by two sets of F4s carrying 750-pound high-drags and napalm. I was very impressed with the accuracy of the pilots operating under such restrictive and hazardous conditions. Visibility was reducing all the time as more and more smoke filled the air. Very soon the fighters had expended all their ordnance and we had completely ringed the beleaguered unit with protective ordnance.

'Raven 66, this is Tamale 35. Is that about what you wanted?'

'Real fine, Tamale. We owe you one, thanks a lot.'

'OK, 66. I will be hanging about for a while. Give me a call if you require any more help.'

'Thanks, Tamale.'

I had about 30 minutes of safe fuel remaining so I climbed up to 2000 feet and held five kilometres to the south, out of harm's way, listening to the commotion going on as the battle continued.

After about fifteen minutes there was no recall from Raven 66 so I headed for Tan An. I had been on the go for 18 hours and

THE 'Y' BRIDGE IN SAIGON

I was feeling very weary. By the time I got back to Tan An and commenced my approached I was starting to feel relaxed—too relaxed! It was customary not to use the brakes until getting towards the end of the landing run at slow speed. When the aircraft was about 200 metres from the end of the runway I applied the brakes but only the right one worked. As the dark void at the runway's end with a minefield beyond loomed nearer I was still travelling at about 40 knots (75 kilometres per hour). I was too close to the end to go around again so I had to kick on the good brake and cause the O-1 to spin around on one wheel in a ground loop. Fortunately the undercarriage did not give way and I skated to a halt with only metres to spare. I jumped out quickly as the crew chief arrived saying, 'Shit man!' We pushed the aircraft into the revetment near the end of the runway. The problem with the left brake was soon discovered: I had taken a 30-calibre bullet through the left brake assembly.

This has been some day. By the time I got to bed at 1.30 a.m. I had been up for 19½ hours. During this time I directed 12 fighters, dropped 33 tonnes of bombs, 6 tonnes of napalm as well as delivered a good load of CBU and 20 mm cannon fire. I had been airborne for 12½ hours, and been shot at for maybe 5 hours. The mission started badly when I was almost killed in a thunderstorm. Finally I suffered the indignation of having to ground loop my aircraft to prevent it disappearing into a minefield. However, the final results were worth it. I know that we had been able to help save many American lives. That is the most satisfying result for me when on these missions. The official tally was 21 structures destroyed, 10 buildings demolished, 30 bunkers blown up and 230 enemy killed.

I was looking forward to some very serious sack time, but at 5 a.m. I was woken again. Captain Don Washburn was to fly me up to Long Thahn to pick up a temporary replacement aircraft for the one I had damaged the previous night and spare parts to fix the grounded aircraft. Don was due on target by 6 a.m. and as there were no brake spares at Long Thahn, I had to fly the replacement aircraft on to Bien Hoa to pick up parts. On the way back

SOCK IT TO 'EM BABY

I skirted around the Y Bridge area where the contact was still raging. I checked in with Tamale Control to see if I might be needed but was told to get the aircraft back pronto as they had just lost another Cessna O-1. Captain Ike Payne had been shot down just off the eastern end of the runway at Tan An. Normally the prevailing wind had us taking off into the west and over open country, where we often took ground fire. To the east it was normally populated with friendly Vietnamese and we preferred taking off in this direction, as we never encountered ground fire. However, this was the return of Tet and who knows what enemy elements may have infiltrated that area. Ike was taking off into the east and turning left at 200 feet having just got airborne, when a burst of small calibre fire hit his engine, stopping it dead. There was also a very strong easterly wind associated with a thunderstorm blowing and we believe Ike got caught in a micro-burst, which drove him straight through a grass palm-woven house, killing four civilians inside, including two children. Ike was severely injured and had been medivaced to Long Binh by the time I arrived back. I had previously had many long and meaningful discussions with Ike and I was very sad to see him go this way.

As there were already two pilots on station at the Y Bridge, Major Bill Walker and Captain Andy Anderson, I headed back to Tan An with the spare brake assembly. Within a half hour on the ground we had two aircraft armed and serviceable. Don Washburn and myself were placed on alert to replace Walker and Anderson if required. Due to the distance of the base from the airstrip, and the associated dangers of travelling the road, we had a hooch with four bunks made available to us right next to the flight line on the edge of the hospital complex. Both Don and myself tried to get some rest in the sweltering heat of the day; however, it was impossible with the noise of aircraft and helicopters operating so near us. The artillery constantly thundering the bass accompaniment, so all we could do was lay there in a lather of sweat. After two hours we were both scrambled for airborne alert closer to the contact.

I was grateful to be back in the cool air at 2000 feet. We held off the southwest tip of Saigon for three hours but were not

THE 'Y' BRIDGE IN SAIGON

required to place any air strikes. That may seem a waste but had we been required we would have saved an hour getting ordnance onto the enemy. Holding also gave us an insight as to how the battle was progressing. We could switch from channel to channel and listen to the ground activity. There was no point holding for more than three hours as we would then not have enough fuel to be of much use if air strikes were then required.

Don returned to Tan An separately and was refuelling when I arrived. We were both looking forward to some sleep, only to find our temporary hooch had been blown to bits after taking some enemy mortar hits. There was quite a lot of blood among the ruins and we were concerned that Bill or Andy may have become casualties. As it transpired a 2nd Brigade FAC was having his aircraft serviced at Tan An rather than fly all the way to Dong Tam a further 30 minutes away. He decided to take a rest in our hooch and while he was lying there, sweating, the VC lobbed a few mortar rounds on the hospital. He was asleep and was woken with a start when the first rounds hit. When this happens you always automatically take flight for the safety of a bunker. That is exactly what he did but a more sensible action would have been to throw himself under the bunk and wait for the right time to run for the bunker. That is easy to say, and I always acted the way he did when awoken from a deep sleep by incoming ordnance. Unfortunately, he reached the door just as a mortar went off a few metres away. The blast blew him back inside. Immediately another round went through the roof of the hooch and he was blown outside again like a rag doll. He was seriously wounded and sustained multiple shrapnel wounds. By the time I arrived he had been medivaced to Long Binh Hospital, a major facility north of Saigon.

Some time later I was talking with another 2nd Brigade FAC who said his colleague had survived. That was the good news. The big problem with multiple shrapnel wounds is that the medics are sometimes unable to remove all the metal so the patient has to wait for the metal to work its way to the skin surface. In some cases this can take years, or the metal may even remain embedded for life, or the foreign bodies can work their way into vital organs with fatal results. His colleague had received a letter from the wounded FAC while he was receiving treatment in Honolulu. He had jokingly written that he was probably the only FAC sent out to have a compass swing done on him to remove the 'A Error' every twelve months due to the amount a metal still in his body. (Foreign metal has a deleterious effect on aircraft compasses. It produces an error in its indication called 'A Error' and has to be periodically adjusted for accuracy.)

Fortunately it was not raining so Don and I resurrected a couple of undamaged bunks in the shattered hooch and bedded down. Two hours later we were scrambled again for Saigon. Our target was the Cholon District of South Saigon. Don was to work targets south of the Kihn Doi Canal while my targets were at the northern entry to the Y Bridge itself. It was rather exciting flying up and down the streets and boulevards of Saigon city at low altitude and firing Willy Petes through the windows of two- and three-storey buildings. By the time we started working with the fighters it was already getting dark. The scene was like Armageddon. There were a lot of fires and smoke, making it hard to see clearly in the failing light. The sky was leaden and criss-crossed by tracers flying in all directions. The radios were jammed with excited voices and the usual radio courtesies could not be shown. To be heard, it was a matter of transmitting over other people and hoping our elevation would make our signal strong enough to be received. This was particularly so on FM. Fortunately on UHF we could usually find a relatively quiet frequency to work the fighters on.

My first set of fighters were two Huns from the 614th Tactical Fighter Squadron led by Lieutenant Colonel Joe Turner. There was no time for catch-up banter and we quickly got down to work. We were forced to use the canal as a barrier between the two major contacts because both sides of the Kihn Doi Canal were engaged in battle. This meant I had to run my fighters parallel with the canal in a west to east direction and break left over Saigon city while the fighters on the south side of the canal did

THE 'Y' BRIDGE IN SAIGON

the reverse. Once again the airliners, Boeing 707s, 727s and DC8s, flew through our flight paths with artillery impacting among us from all directions. We simply had to watch out for the airliners and ignore the artillery.

With weapons delivery the fighters generally liked to get rid of the napalm first for two reasons: the first was that the napalm canisters were heavy and this interfered with aircraft manoeuvrability, and the second was to avoid having this highly volatile weapon being ignited by ground fire while on the aircraft. That would be a disaster. With this in mind I had Joe Turner lay down the napalm first. My mark had to be just right as the napalm had to finish short of the bridge because there were friendly troops on the bridge itself. Fortunately the buildings and structures tended to restrict the spread of napalm and funnelled it like a stream of lava. More by luck than good management, we did not set fire to the bridge. I kept myself tucked in really close to the exploding ordnance, not only to view its effects, but also to look out for elements of friendly troops.

Despite the briefing on FM from the ground commander as to the distribution of friendly troops, I kept picking up groups of GIs in areas where they were thought not to be. Some were alarmingly close to where we were bombing. It must have been very hard to keep track of all the troops during street fighting. This type of fighting along with enemy snipers caused big problems for the US Army. At this stage I was working with Bravo Company, 2nd Battalion, 47th Infantry (opcon to the 3rd Battalion, 39th Infantry) when their Commander, Captain James B. Craig was hit in the chest by sniper fire and had to be medivaced.

As it became darker the fighters were experiencing problems making visual contact with me due to the amount of air traffic and other lights in the air and on the ground. At times I was forced to turn on my landing lights until the fighter had released his ordnance. This action not only made me visible to the fighters but also a target for the enemy. I was attracting more ground fire than I was comfortable with. As I later learned, any ground fire I received was quickly responded to by the GIs who were in pockets all over the place among the enemy elements. It was a very confused and complex ground situation.

At the time I did not realise that this high-combat intensity was not the usual long-term situation. I was quite concerned as to my physical and mental stamina to keep this up for any length of time. During this contact my physical and mental powers were being taxed to the limit. I realised I was at the edge of my capability. Each air strike with troops-in-contact required a high degree of physical and mental effort which was achievable only with infrequent demands. When the same effort is required day in and day out over a long period, the body breaks down. I never had problems getting off to sleep after combat, because I would fall asleep from sheer exhaustion. This sleep, however, was not relaxed or beneficial, particularly in the steamy, tropical climate, so I would always wake-up feeling wrung out. Andy Anderson could never sleep without consuming half a bottle of Bourbon. His favourite was Canadian Club, 'with as little water as possible', he would say.

Throughout this period we found ourselves sharing aircraft with the 1st and 2nd Brigades. At times one could arrive at the flight line only to find there was not an aircraft there but one would soon turn up, as there were about ten on strength with the three brigades. On one occasion I found an empty tarmac so Tamale Control were busy finding an aircraft for me. They contacted a 2nd Brigade FAC on his way back to Dong Tam and he obligingly landed at Tan An so I could use his aircraft. It also meant that he didn't have to fly the extra 30 minutes to Dong Tam so he could get some extra rest. And, as he was out of his home base he could not be used on other tasks so could rest until I got back some three to four hours later. As the aircraft was being refuelled I did a quick pre-flight inspection as I always thought it wise to do my own inspections. I found a neat eight-inch round hole through the tail-plane. A friendly artillery shell on its way to the target had hit this aircraft. The projectile had been a dud, otherwise it would have blown the aircraft to pieces. Had it had VT

THE 'Y' BRIDGE IN SAIGON

fusing it could have exploded within a few metres of the aircraft. When I pointed this out to the FAC he went a deathly shade of white, as he had not noticed the impact of the shell with all the noise and manoeuvring going on.

By late evening I had directed eight Huns and two A37s with each aircraft carrying 750 high-drag bombs, napalm, CBU and 20 mm cannons. I was totally stuffed when I finally touched down at Tan An in the dark.

6

THE MAY OFFENSIVE CONTINUED

HAVING GOT TO BED at about 1 a.m. on 12 May I was thankful not to be bothered until mid morning, when I was awoken by Andy Anderson. I was required to ride back seat with him on a VR over the southern approaches to Saigon. Four eyes were better than two and it was decided that the workload was too high to have pilots flying alone, particularly at night. However, most of the time there were not enough pilots available and we found operating with two pilots was a seldom enjoyed luxury. We spent three hours circling around areas from Tan An to Saigon, listening out on all our radios, but we were not required so we returned to Tan An. After the past few days it was pleasant not to be involved in some sort of combat and just relax in the back seat and enjoy the cool air.

In between flying I was doing nothing except trying to rest as by now we never knew when we would be required to fly. I had left the habit of eating regular meals in the past and now only bothered to eat when there was nothing better to do. My stomach had shrunk to such an extent that I never even felt hungry. As the evening mealtime approached I didn't even care about going to the chow hall as I knew I would be disappointed

THE MAY OFFENSIVE CONTINUED

in the quality of the food. There were always tins of cream corn and biscuits on hand, so I indulged in these tasty morsels and lay down on my bunk.

I was just starting to feel relaxed when I was scrambled to assist units in contact with the enemy at XS 842 883, in the Y Bridge area again. The first fighters I received were two F4s carrying 750-pound high-drag bombs and napalm. The F4s were followed by a pair of F100s carrying 500-pound slicks and napalm. The smoke from all the fires in Saigon was keeping visibility poor and, with the heavy overcast sky and failing daylight, I had to get in low and close to the contact to see. Although we were trained to keep above 1500 feet, the small arms envelope, I was finding it safer and more effective to get down low, particularly at night, as the enemy had only fleeting glimpses of you and could not see you long enough to track you for effective aiming. The downside of this was the closeness of the exploding ordnance, the helicopter traffic, or simply getting disorientated and flying into the trees. On this particular target the buildings were burning profusely so visual references were good and there was little chance of disorientation. At night, the enemy had to rely upon their ability to put enough ammunition into the air for you to just fly into a lead overcast. It was a matter of weighing one set of circumstances against others and hopefully coming up with the right answer.

I was so close to some of the explosions and napalm that I was conscious of the acrid smell of the spent ordnance in the air and I could feel the concussion of the bombs and heat from the napalm through the open window. As we were bombing brick buildings there was a lot of debris flying into the air. A post-flight inspection revealed a 15-centimetre jagged hole, along with a few smaller holes, in the under surface of the left wing of my aircraft. The damage was either side of the rocket launchers—the projectiles had just missed the flying control wires, wing spars and fuel tanks. With all the noise and violent manoeuvres over the target I could not recall when it happened exactly as I had been so close to bomb blasts on a number of occasions. Inside the larger hole was a piece of bomb casing, which I kept for many years until it was stolen, along with many other of my Vietnam keepsakes, during a violent home invasion in 1976 when I was living in Hong Kong. I was in Djakarta operating a flight to Perth with Cathay Pacific Airways. Eight thugs broke into my home, firing guns and wielding machetes. They beat up my wife and mother-in-law and attacked my father-in-law with a machete as well as chopping up our dog and wrecking the home. It took 86 stitches to reattach my father-in-law's scalp.

With the exception of the damage to my aircraft, I considered this last mission to be fairly routine. A USAF Colonel, John E. Pitts, was close by on the ground observing the air strike and he must have been impressed as he submitted me for a US Distinguished Flying Cross. Perhaps he saw things I did not, although I later learned that the air strikes accounted for 58 enemy soldiers. I did not know about this award until December 1980 as it had fallen into the hands of the Australian Government. By then I was working for Saudia, the national airline of Saudi Arabia, as a Lockheed Tristar captain, when the US Ambassador contacted me and said I should call by his Consul. He advised me that he had a number of US awards that had just been processed and wanted to present mine, for which he had organised an impressive awards ceremony. On the day of the presentation my wife and I were picked up from our home in a chauffeur-driven limousine and driven to the embassy, where there was a red carpet reception. Despite the 40-degree Celsius heat, a fully uniformed marine guard marched us into the reception rooms where the awards were made in the presence of about 100 guests and dignitaries. One invited guest was the Australian ambassador, who appeared quite embarrassed about the whole deal. Here was a foreign power honouring someone whose own government would not. All the guests lined up and shook my hand in turn. The Australian ambassador shook my hand reservedly but did not offer his congratulations.

In 1978, when I was working for Sri Lanka's national airline, a similar presentation of two awards had been made to me at the US Embassy but this was not nearly as lavish as the one in Jeddah. News of the awards spread to Australia where the animosity

THE MAY OFFENSIVE CONTINUED

toward me started among those who were still being denied their foreign awards by the Australian Government. Since 1968 the HQAFV had been intercepting all recommendations of US awards to Australians to ensure they were not processed. This was under the direction of the Australian Government, which did not want Australians receiving foreign awards. In this particular case, it would appear that copies of the recommendations had remained in the US system and were eventually processed without reference to the Australian Government, which still held the original recommendations.

Back at Tan An, Bill Walker and the crew chief were not impressed with the damage and said I must have been getting careless. Jake Johnson added his usual, 'Shit, Man!' My matter-of-fact response annoyed them even further. My attitude was not out of disrespect to them but caused by fatigue. Everyone was getting short on temper and patience due to the around-the-clock action. Even though it was 9 p.m., after establishing that the aircraft was still fit to be ferried, Bill Walker dispatched me up to Bien Hoa for repairs. Although I was dead tired I was looking forward to 'being out of the loop', so to speak, while the aircraft was being repaired at the maintenance base. After delivering the aircraft to maintenance I found an unoccupied bunk in a hooch and, with no pillow, sheets or mosquito net, I was in a deep sleep before midnight. I soon realised the hooch I had found was always unoccupied and I kept it a secret from others and used it frequently on my visits to Bien Hoa. This was my holiday shack.

In the morning I checked with the maintenance officer and found out that they had not yet replaced the wing due to the pressure of other work. The maintenance officer thought I was pushing him to fix the aircraft but, little did he know, I was elated at it not being ready. There was an open invitation to all FACs by the F100 pilots of the 'Buzzard Hooch' (510th Tactical Flying Squadron) to drop by and liberally imbibe, in heroic quantities, the nectar of his choice until rendered incapacitated. As their operation was also a 24-hour one, it did not matter what time you called by, there was always a friendly reception going on. Drinks were free, they had a kitchen in which to cook pizzas, and the barbecue seemed to be on all the time. It was estimated that my aircraft wouldn't be ready till late evening so I decided to visit their hooch. I figured I could get a steak for lunch, a few rum and cokes, six hours of sleep and be ready for work by 9 p.m., when my aircraft was scheduled to be serviceable.

By lunch I was well and truly inebriated and the last thing on my mind was eating. It only took a couple of rum and cokes on my empty stomach before I was feeling no pain. By mid-afternoon, I had seen several fighter pilots come and go and, as each new pilot came back from his mission, he was allocated a new assignment—to 'fill in' the Aussie! Just when I was thinking life was good, Bill Walker tracked me down by phone. He said he had arranged for a replacement aircraft and I was to get airborne asap as there were 9th Infantry Division troops-in-contact with the enemy in Saigon. I said, 'But, Sir, I'm pissed.'

Bill Walker replied, 'Well, Cooper, get un-pissed! You are the only one available and we are relying on you', and he put the phone down, leaving the ball in my court.

The fighter pilots gave me a very stiff black coffee and I took several cans of Coke with me to dilute the alcohol and ward off the effects of dehydration. It is amazing with alcohol, the more you drink, the thirstier you get! I obviously heeded Major Walker's command to become 'un-pissed' as, by the time I got airborne, I felt in full control of my actions and thoughts. I find it incredible what will power can do!

Having received the coordinates of the contact immediately on becoming airborne I headed straight for the location without following the usual route around the east side of Saigon. This took me through Tan Son Nhut circuit pattern areas and across Saigon city but I flew at 200 feet. By keeping a keen eye open, I was of no danger to anyone—except myself perhaps!

On arrival in the target area at XS 796 840, seven klicks off the southwest tip of the city, I had just enough time to check in with the Grunts and establish the battle layout when the first fighter-bombers checked in. A pilot I had seen back at Bien Hoa this

THE MAY OFFENSIVE CONTINUED

morning was flying one of these F100s and he was well aware of my condition before take off.

'Hi there, Tamale. How's it hanging?'

'Apart from having a head like a J. Arthur Rank gong, not too bad!' I replied. After we completed putting in the 750-pound bombs, napalm and strafing the target with 20 mike-mike, I gave the fighters a BDA of four structures and three bunkers destroyed. The lead pilot of the F100s was quite happy with the results and as he left he said, 'Thanks, Tamale. You sure work well fuelled on rum and cokes!'

I radioed back, 'Any way of getting another six straight Cokes to me, I'm still bloody thirsty?'

'Ha, Ha, Ha', came the reply.

The contact with the enemy continued into the night and the Grunts were still pinned down so further fighters were ordered. I ended up putting in another five pairs of F100s carrying 750 high-drag bombs, napalm and 20 mike-mike. During the third air strike the Grunts credited us with killing eleven of the enemy. After clearing with the Grunts, who now seemed to be in control of the ground situation, I headed for base with my fuel tanks almost dry. My flying log-book shows two hours of night flying this day so I must have landed back at Tan An about 8 p.m. By this stage I had one hell of a delayed hangover, not surprising with about six rum and cokes, six straight Cokes, one coffee and one steak inside of me. I was not in an enviable physical state. The thought of my own bunk never seemed so good! I had Jake Johnson drive me from the airstrip back to the base. Jake insisted on driving with his lights on full beam—his theory was that the lights would dazzle any would-be attacker. The rest of us used only the parking lights or, preferably, no lights at all, and drove flat out. As the local peasants were not supposed to be out after dark they should not be on the road, so the only hazard then was the occasional water buffalo. At one stage I thought I saw movement off the side of the road ahead of us but I considered it was a reflection from the millions of bird-sized insects Jake's lights were disturbing. Just as we approached that point, my eyes fixed on a couple of forms only three metres off the side of the road. There was a loud clunk as something hit the bonnet of the jeep, bounced through the lowered front windscreen, flew between Jake and myself, rattled around behind the seats and ejected itself out the back of the jeep onto the road and exploded. With Jake repeatedly screaming, 'Shit, man! Shit, man!' I turned off the headlights and slammed my foot on top of Jake's on the accelerator and we sped off into the night. I was aware only of Jake's big white eyes being rather prominent in the darkness. He was not smiling! I believe the grenade was carried forward with our momentum and when it exploded on the road it was very close to where the assailants had thrown it. This would have startled them to the extent that they did not get time to then start firing on us.

Back at base the Army were reluctant to send out any Grunts without knowing the strength of the enemy force. At first light, investigation of the scene showed some blood trails so their own grenade must have wounded some of them. Now they know what 'pissing into the wind' really means! This incident hurried along a yet-to-be-implemented decision for the FACs to have an armed escort whenever they traversed the road by night. The journey from the base to the airstrip did not normally pose a problem in finding an escort, but the reverse trip was all but impossible. After that incident, I always grabbed a bunk in the hospital rather than face the long drawn-out procedure of finding an escort and then putting myself into possible danger by making the trek. Jake spent most nights after this sleeping in the armour plated storage container at the flight-line. So ended another long day.

The next day, 14 May, the gods were smiling on me as I only had one pre-planned air strike to put in. I managed to locate and destroy three structures and six sampans hidden among the waterlilies in a stream at XS 870 738 with three F100s carrying 750 high-drags and 20 mike-mike. This area was less than one klick east of Xom Tan Liem, where I had had the big contact on 10 May. It was probably a landing point for the enemy. The fact that the sampans were still there indicates that not many had escaped our strike. The next day I only had one air strike again,

THE MAY OFFENSIVE CONTINUED

which made me think that perhaps the offensive was over. On the 15 May I flew an escort mission in support of an Army convoy moving from Tan An to An Nhut Tan. The convoy received some sniper fire from a nipa-palm covered area. I requested an air strike of two F100s with 750-pound bombs and we destroyed four bunkers with no return ground fire at XS 650 686, which was on the banks of the Vam Co Dong River, eleven klicks northeast of Tan An.

I was flying my usual O-1 now, tail number 14981, which had been repaired and brought back to Tan An by another pilot. With the beating she had been taking lately, and yet still getting me back home, I started to hold sentimental feelings towards her. She was beginning to have the same effect on me as a faithful pet. In June 2000 I was in Fort Walton Beach, Florida, attending an FAC reunion. Someone there mentioned that the US Navy Aviation Museum at nearby Pensacola was worth a visit. They told me there was a Vietnamese Air Force registered Cessna O-1E hanging from the ceiling with the tail number 5L14981. And there she was, the little lady '981 that had carried me faithfully through dangerous skies some 32 years earlier! Beneath her was a printed history of how, and why, this Vietnamese Air Force aircraft came to be in a US Navy museum.

'My' Cessna was manufactured on 21 September 1951 by Pawnee as an L-19A and was initially assigned to the National Guard in the USA—although the serial number was 51-4981, the USAF-allocated tail number was O-14981. It subsequently served as a FAC aircraft with the 19th TASS at Bien Hoa in the Republic of Vietnam, being allocated to the 3rd Brigade 9th Infantry Division US Army for FAC duties, which is when I came into contact with it. Early in 1975 there were mass evacuations from South Vietnam as the North Vietnamese, against the peace agreements, invaded the south. On 29 April 1975, to escape the besieged city, Major Bung Ly loaded his family of seven into the two seats of '981 at Tan Son Nhut Air Base in Saigon and headed for Con Son Island, off the southeast coast of Vietnam. On arrival, Bung Ly found the place in complete chaos with political prisoners from the infamous Con Son prison rioting. There were hundreds of aircraft using Con Son as a staging point for escape to Thailand. Fuel was not available and, being basically the last aircraft to take off, Bung Ly followed the mass. Without a radio headset or maps, Bung Ly soon fell behind and lost contact with the preceding swarm of aircraft. Out to sea and lost with his family of seven on board, one can only imagine the apprehension Bung Ly was suffering. Low on fuel and ideas, Bung Ly stumbled upon the USS *Midway*. In many documentaries covering the fall of Saigon there is a scene showing the Cessna landing on board the aircraft carrier USS *Midway*. I have seen this footage many times but did not realise until June 2000 that the aircraft was the one that had served me so well.

I met with Major Bung Ly in July 2000 in Orlando, Florida, where he was working at a restaurant. He explained how he hastily scribbled several notes and, on low passes along the deck of the carrier, dropped them explaining his plight. The captain of the *Midway* ordered several millions of dollars worth of Hueys and Chinooks dumped over the side to allow the O-1 to land. The landing was successful and the crew of the *Midway* subsequently sponsored the Bung Ly family in the United States. The Cessna went to the US Navy's Aviation Museum, where she is now safely tucked away in a pristine environment instead of rotting in some junkyard. As I was writing this book, Hurricane Ivan hit Pensacola on 16 September 2004, killing 45 people and causing extensive damage to the Pensacola Naval Air Station. The Aviation Museum was not totally destroyed and '981 survived again. That pleased me! The Birddog with nine lives has finally retired.

On 16 May I was not tasked for any flying so I spent a leisurely day developing my suntan and working on the map system I was creating as the navigation officer. The normal 1:50 000-scaled charts we used to identify targets measured 56 by 68 centimetres—which were awkward to handle in a small cockpit while the slipstream beats in. These charts also soon became unreadable, as they were saturated in sweat and refolded many times and had chinagraph reference marks obliterating the print. My solution was

THE MAY OFFENSIVE CONTINUED

to cut the charts into four sections and insert them into plastic pages in a folder. This meant the pilot could write on the plastic insert, then after each air strike rub it out so they could start with a 'fresh' chart on each subsequent mission. This would eliminated much confusion and the many errors that were made when identifying targets. On the 1:250 000-scaled charts which we used to navigate from point to point, I marked the extremities of each 1:50 000 chart on them. When a pilot was directed to certain target co-ordinates, he would mark this on the quarter page of his 1:50 000 chart and navigated there using his 1:250 000 chart. Major Walker was tickled pink with this concept and it gave me the enthusiasm to complete a set of charts for each pilot.

I was sure the intense activities we had experienced over the previous couple of weeks were over and at 5 p.m. that evening I even visited the chow hall for a meal and was planning an early night. The food was served on stainless steel trays that had indents pressed into them to contain it, thus eliminating the need for plates. Unfortunately these indents were not very deep. We would line up in the chow line and file past the kitchen server, where the cook would dump the meal into the various indents. Because this was not done with any finesse, the gravy would overlap into the ice cream and any fluid food would soon spread from indent to indent. By the time you sat at the meal table, the colourful display in front of you was far from appetising.

Just as I sat down to my rather disgusting tray of bilious gunk, Sergeant Cover called out from the door, as he was not permitted inside the officer's meal hall. Lucky him! 'Captain Cooper, Sir! Troops-in-contact near Saigon—pick you up in five minutes.'

Thank God! I thought to myself. Saved from the cook again!

I left the meal table immediately and jumped into the jeep with Sergeant Cover. He briefed me on the way to the airstrip. Sitting with Cover was a US Army second lieutenant, who was coming along with me as an observer. He was a forward observer from the unit we were about to assist. He had been wounded previously and was now on light duties. He said he had been 'out of the field' for over a month and was keen to be released back to combat.

Contact this time was at XS 822 807, which was one klick southeast of Xom Tan Liem, so it appeared that elements of the NVA were still there but retreating to the south to join the waterways leading back to the Cambodian border's Parrots Beak and the Ho Chi Minh Trail. By the time I arrived on target it was dark. My favourite fighting environment! The only lights I could see were those of Saigon, five klicks to the north. Tonight there were no tracers or other telltale signs of any action. Below me was a matt of inky blackness with nothing on the ground discernable. I switched to the appropriate FM frequency and called the commander on the ground.

'Sandal Alpha 6, Sandal Alpha 6, this is Tamale 35.' Almost instantly a hushed voice crackled in my eye phones.

'Tamale 35, this is Sandal Alpha 6. I have two units pinned down. Each time we try to move we get zapped. We're surrounded.'

'Alpha 6, how spread out are your units?'

'Ah, Tamale, we are all within 50 metres.'

'Alright, Alpha 6. Have someone at the centre of your spread set his strobe going.'

'OK, Tamale.'

In the inky darkness below a solitary, lonely strobe light blinked in the sea of black as much to say, 'Here I am, come and save me!'

'OK, Alpha 6, I have your strobe—standby.'

Like clockwork, Moonshine and the fighters checked in almost simultaneously. I set up a tight orbit over the position where I had noted the strobe on the ground and had Moonshine start throwing out flares directly above me to illuminate the area. There was no wind and the parachute flares hovered past me like dancing, burning ghosts in the night. This proved to be a good first choice as I did not have to correct the drop point for the rest of the mission. Then I briefed the fighters.

When illuminated I could see that the contact was close to the western banks of the Rach Gieu. All the streams in this area linked together, allowing passage by water all the way to Cambodia. There was still no tracer fire coming my way and my thoughts

THE MAY OFFENSIVE CONTINUED

were that the enemy did not want to compromise their position and were waiting to see if I could locate them.

'Tamale, this is Alpha 6.'

'Go, Alpha 6.'

'Tamale, we hear some AK47 gunfire but it is not coming our way—it must be directed at you.'

'Thanks, Alpha 6.'

I had been looking for tracer fire but the enemy were not using tracers. As soon as the flares started to illuminate, the enemy started firing and I was now picking up muzzle flashes from a number of guns firing directly at me. This meant that the ammunition from the muzzle flashes I could see was going behind me. There was no point worrying about the flashes I could not see. I just had to get on with the job. The area where the muzzle flashes were coming from would get my attention first.

The second lieutenant in the back seat was young and keen. He appeared unfazed by his recent wounds and was constantly on the intercom asking me questions. He kept me briefed on who I was talking to on the ground by saying such things as, 'That's Bob', or, 'That's Jo', and so on. This was a little annoying when I was so busy talking with so many different agencies, however, I did not have the heart to dampen his immature enthusiasm by telling him to shut up.

Time seems to fly when you get so involved with a task, such as directing air strikes, and before I knew it I had directed two flights of F4s and one flight of F100s. I had been airborne for 3 hours and 40 minutes and my fuel was getting low. As the grunts were still unable to move, I ordered a replacement FAC and more air strikes during my third attack. About this time the back seater had gone quiet. He didn't answer when I tried to contact him on the intercom. When I looked over my shoulder he appeared to be asleep but on shining my flashlight on him I could see blood running down his neck from a hole in his helmet. I could not stop putting in the air strikes as he was only one, on the ground there were many.

'Tamale Control, this is 35.'

'This is Control, I have been monitoring. Tamale 32 is on his way and I have ordered more fighters.' What would we do without Sergeant Cover! Worth his weight in gold!

'Alpha 6, this is Tamale 35.'

'Go, 35.'

'I have done all I can for now—sit tight—you should have Tamale 32 with you in 30 minutes plus more fighters—I have to leave straight away as Alpha Deuce Six (26) has taken one in the head.'

'Christ, Tamale. I was there when they medivaced him a few weeks ago—look after him—he is a good man.'

'Roger, Alpha 6—stay safe!'

I switched to VHF. 'Tamale Control, this is 35, my back seater is WIA—require a medic on arrival Tan An.'

'Roger, Tamale.'

'Tamale 35, this is Tamale 32. Go ahead with your briefing.' I briefed Andy Anderson on the target situation and bid him a good evening. Then I flew at full throttle to Tan An, doing a tight turn onto final, virtually touching down during the turn, and taxied at high speed to the waiting ambulance parked at the end of the strip. In seconds the medics had the lieutenant on his way to surgery. A bullet had gone through the left rear window and hit his helmet, taking off his left ear, then exited the rear of his helmet and out the right window. There were no other hits on the aircraft. The lieutenant would be a little longer than he wanted getting back onto combat status.

When Sergeant Cover woke me in the morning advising me I had to go back to last night's target, I noted Andy Anderson was in bed, so he must have had a successful mission. As we arrived at the flight line, Don Washburn was just landing from operating the same target. Don briefed me on how he last saw the battlefield as I strapped into my aircraft, '981 again. On arrival at the contact the grunts sounded distinctly tired. This was their second day at it. How glad I was to be in the Air Force! While the FAC's quarters at forward bases were rather basic to say the least, and not nearly as good as the pilot's quarters on main bases, they were

THE MAY OFFENSIVE CONTINUED

heaven compared with what the grunts had to contend with in the field.

As I approached the target area I could see volumes of ominous black smoke billowing skywards and puffs of white smoke dotting the sky from the artillery markers as the grunts zeroed their artillery on the enemy positions. The black smoke was coming from a downed helicopter that had three occupants still on board, ending their war.

'Alpha 6, this is Tamale 35. Still at it?'

'Roger that, Tamale. We will have to stop meeting like this! The main element of resistance is from that tree line 100 metres north of the downed chopper. We haven't been able to get the bodies out as it is too hot.'

'Got it, Alpha 6. Keep your artillery lobbing in 200 metres north of the tree line to seal the enemy retreat line and I will have the fighters attack from east to west.' By listening to the artillery frequency, I knew they were using a FSB to the north of the contact. Attacking east-west kept the fighters out of the max ord of the artillery fire.

'Roger, Tamale. Could you have the fighters attack west to east to prevent them attacking across my Tracks?'

'Negative, Alpha 6. With the rising sun we will have better target definition attacking east to west.'

'Roger, Tamale. Your ball game.'

I had two pairs of F100s carrying an ideal load for this target of 750-pound high-drags, napalm and 20 mike-mike. We did a complete coverage of the tree line for about 200 metres in length and 50 metres width. With the artillery impacting on the VC escape route to the river, some of the enemy broke into the open trying to escape to another tree line to the south. I was able to direct the fighters onto these. The fighters strafed the enemy as they ran in the open. The fighter pilots were elated as they often did not see the enemy they were attacking.

As we were finishing strafing I noticed the Tracks moving in from the east in order to get in on some of the kill. I could hear the relief in the ground commander's voice as he radioed, 'Damn good job, Tamale. We can take it from here.'

I climbed up to 4000 feet and flew cover for 30 minutes. I watched the grunts moving through the area devastated by our bombing and surrounding areas. Some of the Tracks pulled up near the chopper, which was still burning fiercely. As the Army were in control I headed back to Tan An after 3 hours 30 minutes in the air.

The next day was 18 May and I was required to attend a debriefing at the fire support base (FSB) at Binh Phuoc, concerning the enemy contacts over the last two days. High ranking Army officers wanted to know why it had taken so long to suppress the enemy. I was picked up by a 20-year-old Army lieutenant flying an OH23 Raven and I believe I was more scared with him than I ever was in combat. He did not fly above 50 feet. Anything higher than that, he went around and not over. I was expecting at any time to be confronted with a bunch of VC at eye level. On arriving at Binh Phuoc the FSB were firing out to the southwest, which was directly across the chopper landing zone (LZ). Instead of holding off until the firing stopped, he flew right under the guns and landed with the guns firing about 100 metres from us. We were looking straight down the barrels and the flashes from the big guns were highly visible, even in the daylight. We departed the same way—me clutching my seat with white knuckles. Knowing artillery trajectories probably made this operation quite safe but it was still very unnerving from a passenger's point of view. It was a great relief to get back to flying my own aircraft, where I had some measure of control over my destiny.

In the afternoon I flew '981 up to Bien Hoa for some routine maintenance. The maintenance was scheduled to take four hours so I slid across to the fighter pilot's hooch for some light relief, but I wasn't drinking today! I ran into one of the pilots who was on one of the air strikes yesterday. He showed me his gun camera film that clearly showed the 20 mike-mike tearing up the ground among a number of VC in the open. That film was going into the squadron archives as a collector's piece.

I discussed the number of hits and near misses I had had in the last month with a USAF major. His solution was, 'Get a lucky

THE MAY OFFENSIVE CONTINUED

charm!' Many of the fighter pilots carried a rabbit's foot, dollar piece, dice or some other trinket that they found comfort in. Although I never believed in that sort of thing, I had nothing to lose by getting one. I headed down to the base exchange (BX) to see what I could find. There were numerous trinkets available but only one took my fancy and I bought it. It was a small silver, oval-shaped locket with a picture of a pretty girl inside. Some two years later someone identified the pretty girl as Olivia Newton John, who was just becoming popular as a singer. I religiously kept this trinket attached to the pocket zipper of my flight suit. There were three occasions later when something nasty happened to me and on all three of these occasions I was not carrying my lucky trinket. As I have said, I never believed in this mumbo-jumbo but that was certainly food for thought. Many years later I was living just down the road from Olivia so I gave her the locket after having had it all those years. At the time she was suffering from cancer and I hoped the trinket would aid her recovery. Each time I have met her since I have forgotten to enquire about the fate of my lucky charm.

On the flight back from Bien Hoa, I saw a formation of Hueys inserting troops just near Tan An so I decided to circle the operation from a vantage point out of the way. The departure from the insertion did not appear to be orderly and one of the helicopters was trailing smoke. Some of the Hueys put down not far from where they had just departed. They were obviously in trouble so I asked Tamale Control for their frequency. Control advised that it was an ARVN operation but he would try to come up with a contact. By the time he did, it was too late for me to do anything as I was low on fuel. I later found out that it was the 135th Assault Helicopter Company (AHC), US Army inserting ARVN troops. They had come under heavy ground fire at what was supposed to be a secure LZ. There were ten helicopters and all were hit. Not many made it back to their base, crashing on the way. Two Hueys were destroyed and the other eight badly damaged. Interestingly, I was to learn that many of the crews on this operation were Royal Australian Navy.

On 19 May I only had one pre-planned air strike with three F100s in the 'Bowling Green' area at XS 646 695. We destroyed two structures and four bunkers with no sign of the enemy. Back at Tan An, Major Walker walked into the hooch where we were all idly laying around and announced, 'Listen up, you guys!' We gathered around him as he continued.

'The 9th are having a ceremony tomorrow at the Y Bridge. We are all going to stand knee-deep in bomb craters, blow the bugle and wave flags—good fun. And all us victorious heroes are going to get medals. Andy and Don are to get DFCs, Cooper and myself, Silver Stars and Mr Cover, the Bronze Star. Also, we are all getting Army Commendation Medals for our time in the TOC between missions.'

A big to-do was planned with generals from all services present, a band and the press—it was to be an historic event. Major Walker asked for details of my Australian commander in order for him to be advised in case they wanted to get in on the publicity surrounding the only Australian to take part in the air battle. Within two hours a message came back from Air Commodore Dowling saying, 'Under no circumstances is Flight Lieutenant Cooper to accept the Silver Star nor is he to appear on the parade scheduled for 20 May.'

Major Walker was dumbfounded and got on the phone to the Australian Embassy only to be advised the message was correct. Bill Walker contacted the office of the commander of the 9th Infantry Division, Major General Ewell, who also took up the matter. The Australians rudely ignored his contact. Bill Walker said, 'God damn it! What do the Australians find wrong with our awards? God damn, I don't believe this shit!' On 2 July 1969 General Ewell, by then a three-star lieutenant general, wrote to Major General R.A. Hay, Commander Australian Forces Vietnam, enquiring again about the situation regarding US awards to Australians. There is no record of a reply on file.

The Americans were all insulted by the Australian side's attitude. The only Australian to attend the parade was Major General E.L. MacDonald, the Australian Forces Commander, as a

THE MAY OFFENSIVE CONTINUED

guest. I was also rapidly becoming disenchanted with the disgraceful conduct of my chain of command. On the 20 May my first son had been born but the Australians did not let me know for seven days. It had taken the RAAF commander two hours to prevent me receiving recognition from the Americans but seven days to let me know about the birth of my son. While all my flight colleagues were at the awards ceremony, I was covering our AO by myself. This lone Australian was 'holding down the fort' for the Americans. I did a total 6 hours, 25 minutes in the air, controlled 14 F100s with seven air strikes, sunk six sampans, destroyed two structures and 12 bunkers. All this on the day I was supposed to be receiving recognition for gallantry and celebrating the birth of a son. To add insult to injury, I also flew to Long Thanh North to pick up Major Walker after the awards ceremony. Sensing my disappointment, he did not discuss the ceremony with me, or show me his medal, although I did see him showing his Silver Star to others. At least, someone had a heart! On the positive side, I had the AO all to myself—it was mine for the day!

The following day I had a pre-planned air strike at XS 582 784, just five klicks northwest of the Ben Luc bridge. Intelligence had evidence of VC activity in the area. This evidence was usually reports from locals hoping to pick up some payment for their information. Many reports were fabrications but all of them had to be followed up unless they were known to be incorrect. I approached the area at 4000 feet in order not to broadcast my presence. The lay of the land was unusual in that it reminded me of the remote swamplands of New Guinea. There was low, bushy vegetation along with the reflection of water. This ceased abruptly at the start of a line of very tall, but sparsely situated trees. It was the tree line where I concentrated my search for any signs of the enemy. As I was circling I saw about 50 soldiers to the south of the tree line, which they were yet to penetrate. At first I thought they were VC moving in the open in daylight, which was unusual. Focusing my binoculars, I could make out more clearly that they were ARVN troops. We did not often receive warning of where ARVN troops were as they did not advise us of their movements. My fighters were not due on target for another 15 minutes so I searched the tree line at the point where I thought the ARVN troops would penetrate the woods. Just when I'd decided that I would put in an air strike in the tall timber, something caught my eye about 100 metres north of the tree line. I trained my binoculars at the sea of swamp and saw there was a natural area of raised land above the water level. A slight discolouration of the vegetation indicated a recent disturbance. Among the bushes were about a dozen VC. Plotting the direction the ARVN troops were taking, it seemed to me that they were walking into an ambush.

I tried contacting the ARVN troops on the ground but could not get an appropriate frequency. I would only compromise their position if I flew low over them so I considered firing Willie Petes close to their lead troops to try to halt them so they would check to see why I was firing at them, or do something sensible. While I pondered this dilemma my fighters checked in so I quickly briefed the two F100s and told them to hold at 10 000 feet over the Ben Luc bridge, a little distance from the enemy, so as not to alert them. My intention was to surprise the enemy, who seemed confidently settled into their ambush position. The fighters had 500-pound bombs, which could be released at a higher angle and altitude than the high-drags—just what I needed in this situation. They were also carrying napalm, CBU and 20 mm cannons. I told the fighters I would roll in for the mark and they were clear to salvo their 500-pound bombs on my smoke without further clearance. I gave them random headings, which they liked, but this meant I had to keep them in sight at all times. I closed my throttle, rolled on my back and pulled my sighting point through to the target. With no engine noise the VC would only hear the whooshing of my slipstream at best. My marker hit right among the enemy, who started to scatter in panic. With the random heading clearance the fighters were right on top of them with some beautifully placed bombs. I then cleared the fighters to use their bomb craters as the centre from where I wanted them to spread, and to expend the rest of their ordnance. For ten minutes I sat

THE MAY OFFENSIVE CONTINUED

directly over the target at 500 feet and monitored the fighters as they flew underneath me, destroying the countryside. After the fighters had left I flew low over the ARVN troops who had gone to ground. They were all waving in a rather enthusiastic manner so I gathered they comprehended what had transpired. A few days later we received a credit of six VC killed. I am sure there were more but the ARVN needed a body count too!

The next morning, due to what I had found on the previous day, I was given another air strike back at the same coordinates. Concentrating on the tree line nearest where I had surprised the VC, I destroyed five bunkers using two F100s. No sooner had I cleared the fighters off that target than Tamale Control called me to attend to a troops-in-contact at XS 725 710. This was only one klick northwest of Rach Kien and 17 klicks from my position. By the time I got there, two A37s were checking in with me. They had good loiter time available, so I was able to have them hold while I sorted out the ground situation with the grunts. It was not a complicated situation. Using 500-pound slicks, napalm and CBU, the enemy were soon quietened and the grunts were up and moving again.

My flying logbook shows that I flew to Bien Hoa after this air strike. I cannot remember why but it was probably to get an oil change done to the aircraft. By evening I was back at Tan An. I had not flown at night for five days now and I think I could comfortably get used to this daylight flying!

On 23 May I put in three air strikes on the one 2-hour 30-minute flight, all in different areas. The first one was near 'The Testicles'. No sooner had I put that one in than I received a call to put in another five klicks southeast, where an estimated 100 VC were in bunkers. I found obvious evidence of human activity in a remote area and located some bunker entrances, which I bombed with two A37s carrying wall-to-wall 250- and 500-pound bombs with a 100-millisecond delay. This means that the bombs would not explode on the surface but penetrate the ground first, which is ideal for unearthing bunkers. I did not get any return ground fire or reaction so I could see no purpose in expending more ordnance on that target. I then directed an Australian Canberra onto a target closer to Binh Phuoc. After the lively activity of directing fighters, it was always a relaxed situation using the Australian Canberra. Almost boring in fact, but those 1000-pound bombs sure made a nice big hole.

When I arrived back at Tan An Colonel Benson, the 3rd Brigade commander, was just leaving to inspect the results of the air strike I had placed on the suspected 100 VC in bunkers. He invited me along. Just for a different experience and out of curiosity I went. I did feel rather apprehensive walking around the bombed area in the open, away from the relative safety of the base. The Army did this all the time and thought nothing of it. On the other side of the coin, I could never get a grunt in my aircraft voluntarily. They said it was too dangerous! To each his own! It was quite a shock to see close-up the devastation our air strikes made. The ground was so churned up it was difficult to walk over. The acrid smell of the burnt ordnance, vegetation and flesh was an aroma that I will never forget. Body parts were strewn about all over the place—most of it unrecognisable as human. At one stage I trod on something soft, which turned out to be a rib cage. The intestines oozed out of it, making me almost throw up. There were 12 heads counted on the surface and one of the officers said, 'There you go, Cooper. Twelve KBA for you—probably 100 more underground but we 'ain't digging 'em up.' The other grunts did a token clean-up by throwing most of the significantly sized body parts together in a bomb crater. The VC always returned at a later stage to recover their dead.

This nauseating experience upset me quite a lot, but not as much as seeing an American badly injured. I suppose, rationalising, an injured American could well be me but a VC is the enemy and it therefore disassociated from my own feelings. The revetments for our aircraft were right alongside where the dust-off choppers landed the wounded for transport to hospital. One day I saw a blond 20-year-old laying face down on the stretcher. They were taking him to the meat wagon (ambulance). He was covered in mud with no shirt on. His left arm was missing along

THE MAY OFFENSIVE CONTINUED

with his shoulder blade and half his back. That scene really played on my mind for many years and I can still vividly picture it today.

On 24 May I controlled four pairs of A37s. The A37 was used quite a lot as the evaluation period for that aircraft was ongoing. I liked using them as they carried an exceptional load for their size and were very nimble and accurate with the experienced pilots operating them. The targets for the day were generated on VC sightings and were centred around the Vam Co Dong River, just east of the Bowling Green in the Binh Truong Tay area. This was a definite indication that the NVA were on the retreat and making their way back to the Kinh Bo Bo, which would lead them to the Parrot's Beak and the Ho Chi Minh Trail. We destroyed a total of six bunkers and two structures. This might not seem much material damage for the cost and effort put in but anything denied to the enemy was worthwhile and you never knew if you were going to catch the enemy in the bunkers as we had the previous day.

The next day I put in two air strikes four klicks further west, in the Bowling Green itself. In addition to destroying seven bunkers and two structures, I demolished a brand new VC-built bamboo bridge over a small stream. It had had foliage pulled across it to hide its existence, but the vegetation must have been a couple of days old as it had discoloured a bit, which is what drew my attention to it. It occurred to me that we were actually just behind the enemy, bombing their footprints. I kept requesting air strikes further northwest along the Kinh Bo Bo but the Army had their way of doing things and probably knew better than I did.

For the next two days I was on air liaison officer (ALO) duties. This meant attending Army briefings and flying around at low, low altitude with lunatic teenage chopper pilots. Major Walker had decided that we three FACs should undertake some of his duties for experience at the level of our next rank. I found this thoughtful of him as I had seen many officers promoted outside their level of comfort. There was some suggestion that Bill Walker was also getting a little gun-shy as the ALO task was fraught with danger. Knowing Major Walker, I know this was not the case.

He was a gutsy commander. I jokingly said to him, 'But, Sir, I can't possibly do ALO duties as I have the demanding task as tea club officer.' Without showing any signs of humour, he replied, 'You are quite capable of doing both, Cooper.'

As VC snipers tended to pick out officers or any person dressed unusually, we normally had to wear full grunt gear when acting as ALO and show no evidence of being Air Force. For two days I virtually flew, walked, ran, sat and slept in a right echelon position on Colonel Benson. This is what's called being his right-hand man, I imagine. Anytime he wanted to know anything about TACAIR, I was there to give him information instantaneously. I felt a little like Radar from the TV series *M*A*S*H*.

Back to my real world on 28 May, I was rostered for a VR mission along the Kinh Bo Bo. There was plenty of evidence of the enemy having been there, with the disturbed mud in the canal water. This was noticeable all the way to the Parrot's Beak on the Cambodian border. The NVA elements had retreated, it would seem.

I had nothing scheduled for the following day and Major Walker suggested I go to Bien Hoa and visit Colonel Patrick, who wanted to see me when it was convenient. When I arrived at 19 TASS, Colonel Patrick was out flying so I cooled my heels for a while, talking with some of his staff. In due course Colonel Patrick arrived back saturated in sweat. He was a big man and ageing, so the tropical humidity and flying combat did not suit his disposition. 'Hi ya, Cooper. Nice of you to drop by. Come into my office. Coffee?' The Americans always had black coffee brewing. The coffee pot was always black from lack of cleaning and permanent use. All they did around the clock was to add water and coffee as required. It was a terrible brew but I didn't want to offend so I said, 'Yes, thank you, Sir.'

'Now, Cooper, what is all this horseshit about our medals and your government?' I explained that I was not sure.

'The only thing I can think of is petty jealousy. It is very hard to get any sort of medal in our armed forces and I think people back home who have no chance of getting a medal don't want

THE MAY OFFENSIVE CONTINUED

anyone else to get one either. Ironically, some of those policing this issue wear American medals from Korea.'

'Well fuck 'em,' he said. 'We are going to process our recommendations on you Aussies and you are going to get what you have earned.'

With that, we 'chewed the fat' convivially for a while, until the mounting phone calls and paperwork indicated I should leave. Despite Colonel Patrick's good intentions, he did not realise that his recommendations were being intercepted at HQAFV and were never to reach the American processing stage. A diplomatic direction had been released from Canberra ordering Australian commanders to prevent the processing of all US recommendations for awards on Australians. One method HQAFV used was to ask the 504 TASG to send them all the US recommendations on Australians so that our government could award an equivalent Australian medal, after which the US documentation would be returned for US processing. It turns out the Australians were filing all the US recommendations and not returning them to the US system for processing, then denying they ever existed. This denial still continues today.

Early next morning I put in three air strikes four klicks northwest and northeast of Can Giuoc. There was no ground fire detected but we destroyed six bunkers and six structures hidden away in the nipa-palm along small streams. These would have been staging posts for attacks on the army base at Can Giuoc.

On the last day of May I had two pre-planned air strikes on suspected VC sites two klicks west of Highway 4 near Binh Chanh. I circled the target area for some time before the fighters arrived but could find nothing definite to indicate enemy presence. However, when I put my first smoke marker down I started to receive some light, spasmodic ground fire from a couple of areas nearby. The fighters not only carried the usual ordnance but also RP (rocket projectiles), which was a weapon not frequently used in the Delta. After I released the fighters I flew around the general area for about an hour but before returning to Tan An, I flew some low passes over the bombed out area to get a better assessment of the damage. Mixed among the mud were some human forms. In the hot sun, mud on the bodies dried out quicker than the wet ground, giving the bodies a ghost-like appearance. I could not assess the KBA from the air and had to rely on the Army being able to put a patrol through there before the VC cleaned up.

7

DONG TAM

I STARTED JUNE WITH a rostered VR mission. I thought I would concentrate my time looking around likely NVA retreat routes in the swamp areas southwest of Saigon but before I left Major Walker arrived with a full colonel. Bill took me to one side and said, 'We have this turkey from the Inspector General's office and I have been entertaining all morning. They claim that we have been using excessive force in our attacks. Now I want you to take him up on the VR, find a nice passive target and bomb it softly—get my drift?'

Excessive force? Perhaps our ordnance expenditure was considered too large for the results, perhaps the destruction of real estate was too severe or, perhaps they were trying to save money. When troops are in contact with the enemy, I really don't think it is time to budget ordnance.

I noticed that Bill had been giving me all the odd tasks—I never saw the others doing any duty out of the ordinary. I had discussed this with Andy Anderson, who was much more senior to me with time in the service. Andy replied that, 'The major is being cunning. If one of these oddball duties did not work out to plan, with a USAF officer there would be repercussions, but with a foreigner there would probably be no follow up'. This made sense to me, as I am sure I did not come across at being a master of all trades.

Bill introduced me to the colonel, who was cordial but very obviously a REMF (rear echelon motherfucker), as we called those administrators who would sit in an office creating plans that look good on paper but are really unnecessary hurdles for combatants. Major Walker fussed around the colonel as he strapped him into the back seat of my aircraft—'981 as usual. While adjusting the straps he gently and thoroughly briefed him. He was obviously buttering him up and gave me the occasional wink.

I had a feeling that the enemy I had detected on the 21 May were using a different escape route to the usual one—a point on the Vam Co Dong River that led to the next canal that ran off to join the Bo Bo canal via areas of abandoned pineapple plantations. I headed for this area after take off. It was roughly 23 kilometres west south-west of Saigon. I constantly kept the colonel briefed on what I was doing and why as I wanted to do the best for Bill. Pissing in pockets is against my principles. But one of my failings in life has always been intolerance of bullshit and, worse, not being able to stop myself from letting the perpetrator know it. Even so, I tactfully flew around very sedately. Where the small canal ran into Vam Co Dong River at coordinates XS 563 802, I noticed disturbed mudtrails in the water originating in some nipa palm. I flew down to 200 feet, doing a slow turn across the nipa palm rolling on 60 degrees of bank so I could look vertically down through the trees. The enemy suddenly opened fire with AK47s, which was a really stupid thing for them to do. They probably thought that their position had been compromised. Well, it had now! I rolled back to wings level and went for the deck until I was some distance away over the disused pineapple plantations. I then climbed up and continued away from the target so the enemy would think I was leaving. At the same time I called for fighters and very soon had a pair of F100s and a pair of F4s on station. With the brief view I had had down through the vegetation I could see a mixture of VC and NVA—black pyjamas mixed with khaki uniforms.

By this stage I had forgotten the colonel was in the back seat and was back to my old self. As I turned back toward the target area I picked up two sampans heading along the river, hugging the jungle riverbank 100 metres down stream from the mud source. One of my fighters broadcast, 'One away!' but I did not immediately comprehend what was said—this was terminology used by pilots in North Vietnam for a missile launch. Before I could ask him to say it again, there was a deafening bang against the left side of the aircraft. When the shock passed, I was conscious of the acrid smell of cordite and I could see scenery through small holes in an enormous dent just below the throttle. The front windscreen had a large crack in it. I had a jagged piece of metal right through my left hand, between the thumb and index finger, which virtually locked my hand in a semi-clutched position so I could only use that hand for pushing and pulling the throttle. The colonel in the back seat was trying to say something over the intercom and was poking me in the shoulder so I looked around. He was perfectly all right so I shouted commandingly, 'Shut up!'

Once the shock and terror had abated enough for me to start thinking clearly again, I went about laying down the ordnance in a very aggressive manner. When not on the radio, I recall shouting profanities in a rage. I imagine the colonel wondered where that nice, friendly and informative Australian pilot had gone! Before I laid down the bombs, CBU and napalm I directed one of the fighters onto the sampans using 20 mm cannon. The meaningful shape of the F100 looked awesome as it flew along the river at zero feet, streaming thin black smoke from the burnt JP4 aircraft fuel. The pilot picked up the sampans visually just 300 feet below me and slightly to one side, and emptied his guns into them. It was pretty to watch and made me proud to be associated with such professional and accurate pilots. The sampans disappeared in an eruption of water and were eaten up by the cannon fire, leaving nothing visible on the surface after the fighter had passed. The rest of the ordnance was laid down with very little return fire. After the fighters had left I was able to fly low around the area without attracting any ground fire. The place where I had initially seen the

VC and NVA was burnt bare with nothing recognisable. Then it occurred to me that I was basically alone in a remote area without the Army, or any other help. If I were shot down I would be totally at the mercy of the enemy. I quickly removed this thought from my mind. I checked in with Tamale Control, advising them that I had some damage and would recover to Bien Hoa for maintenance as it was not much further from my position than Tan An. Besides, the hospital at Bien Hoa had nice nurses whereas Tan An only had hairy-assed male medics. Obviously, the gender of the staff at Bien Hoa Hospital did not influence my decision in the least!

All the way to Bien Hoa the colonel said nothing and I actually forgot he was there again. As my thoughts drifted to other things the tenseness in my body slowly dispersed only to be replaced by an overrelaxed and dangerous calm. In combat this is when you are at your most vulnerable. I noted this but decided that I was in no danger. On landing at Bien Hoa I thought I was going to get something of a roasting from the colonel, not only for aggressive flying but also for telling him to shut up. I had thought his silence on the flight to Bien Hoa was brought about by him constructing a devastating debrief. Instead he said, 'Well done, Captain! I wanted to end up in Bien Hoa today anyway, but next time you tell a senior officer to shut up, say please. Nice flying with you.'

He then strode off confidently and with determination. I am sure he probably collapsed at the knees when he got out of sight around the corner of the hanger. After getting a time estimate for the repairs to be completed, I took myself off to the hospital to have my hand looked at. As a non-American I was treated like a celebrity because the nurses were curious about this man from 'down under' with a funny accent. My wounded hand required only six stitches to repair it and by the time they had it cleaned up, it did not look nearly as bad as I first thought it was. Because I was from out of base I was offered a bed in the hospital where I was kept under observation in case I developed shock. Bliss! I had the luxury of being served dinner in bed.

The next morning, after breakfast in bed, I had a leisurely shower and put on my clean flight suit, which had been laundered by one of the nurses. They were lovely, obliging women and their friendly attentiveness, under less than ideal conditions, still stirs me today. It was then that I realised I did not have my lucky charm attached to my flight suit. This had not concerned me very much at the time, but later, looking back on it, I realised that for me there was some significance to it. I hung around the hospital talking with the nurses until I started to feel out of place. The aircraft was not going to be released until late afternoon so I visited the base exchange (BX) and wandered around the base, basically sightseeing and talking with anyone who had the time to stop and chat. It was great to be alive!

The flight back to Tan An was without incident but I was dreading what Major Walker might say to me as the previous day's mission had not gone exactly as he would have wished. As I taxied in just as the light was failing after dusk, I could see Bill sitting in his jeep near the revetments. I smiled and waved to break the ice. On the drive back to the base he said, 'I have been wanting to talk with you, Cooper.'

I thought, here it comes! 'Oh, really,' I said, faking surprise.

'Yes. Since Ike crashed I have been doing all the flying instructional requirement myself. I want you to be the CTIP [combat tactics instructor pilot].' He then outlined my duties, which were basically checking out new pilots when they joined the unit and doing periodical evaluation rides on the other pilots already there. Because I was the last to join the 3rd Brigade, and a foreigner as well, I said, 'Wouldn't one of the USAF guys be more suitable?'

Bill bluntly replied, 'No. I have my reasons and have discussed it with Andy and Don—you're it!'

'Well thank you, Major. I will do my best.'

'I am sure you will, Cooper. Of course, I still want you to hold down the navigation officer's slot and continue your outstanding duties as tea club officer', revealing his dry sense of humour.

Any new pilot that joined us was usually checked out on the Birddog, so my primary task was to teach them how to most effectively control the fighters with the least chance of getting shot down. And, of course, all those million and one other little things, such as recognising foliage discolouration, trails through the mud, and other suspicious activities that would be learned with experience. All I could do was to continuously discuss these aspects with them over time. Although the USAF pilots were officially checked out on the Birddog, it was to a minimum standard. Most of them had never flown a tail-wheel aircraft with a propeller on the front as their training was on tricycle-gear jets. This meant that I invariably had to do some basic training with some of them. But I must say, they all generally handled it extremely well.

In the morning I directed three pre-planned air strikes on targets one kilometre west of Rach Kein, to the southwest of Saigon. The army base there had received incoming mortars from this location in the early hours of the morning so the intelligence officer considered it worth pounding the area a little. With 500-pound, one-second delay slicks, we destroyed two structures and ten bunkers. After the first set of fighters had completed their bombing and the foliage was displaced, we could see six sampans along the bank of a small stream. We destroyed these as well with accurate bombing. There were no local Vietnamese near this site so the sampans must have been associated with the VC. I did wonder if we had caught the VC in the bunkers or whether they had perhaps been sitting in a tree near by, laughing at us killing trees. But, we certainly made a mess of their transport and camping facilities.

At the end of the strike, Tamale Control advised by radio that Major Walker wanted to speak with me. 'Hey, Cooper,' came Bill's distinctive Texan drawl. 'Your commander in Saigon wants to see you.' He sounded quite excited. 'Perhaps they are going to give you a medal. Recover to Tan Son Nhut, see your commander, and get back here with the aircraft pronto.' This was very good of Major Walker as it meant a valuable aircraft would be unavailable to them for several hours.

At that stage of the war Tan Son Nhut was having 600 movements an hour and the tower frequency was literally jammed with people trying to get clearance to take off or land. There were light aircraft, airliners, transports and helicopters all continuously on the move. There were also fighters requiring gun-safety pins to be removed while they waited on the runway. I really had to admire the air traffic controllers who managed to keep their calm the whole time. At one stage a formation of Vietnamese Skyraiders, who became tired of trying to obtain take off clearance, just rolled onto the runway and took off, causing a Pan Am Boeing 707 to abort its landing approach. The Skyraiders were fully loaded with bombs. The number four of the formation started an early turn to cut the corner and catch up with the rest of the formation—this is standard practice—but he rolled on too much bank angle and fell out of the sky from 500 feet (150 metres) with an enormous explosion in the rice paddies to the southwest of the airfield. During the brief pause on the radio, as everyone's attention was drawn to the spectacle, a couple of opportunists jumped in and requested take off clearance. Despite the burning fighter off the end of the runway, aircraft continued to take off and land. There was a war going on.

After thirty minutes of circling on downwind leg, dodging high and low speed traffic, I was getting extremely low on fuel—too low to consider going anywhere else to land. On the east side of the runways was a very large tarmac area with C130s and C123s parked on each side. Down the middle looked good to me so I landed there, as there were no transports taxiing in or out. I did a precautionary approach on stall speed, touched down on the edge of the asphalt, rolled no more than 60 metres and I pulled up by the first revetment. On the appearance of this Birddog running along the tarmac a few guys on bicycles wobbled to a standstill. Apart from that, I was surprised how little attention I received—almost disappointed really. I found a parking spot that no-one seemed to object at my using, and secured '981.

I had no idea where to get transport and I could not find a phone number for Air Commodore Dowling. I made my way to 7th AFHQ where I was lucky to stumble on the USAF colonel who had arranged transport for Roger, Mac and myself when we first arrived in Saigon. Once again I had that insecure feeling as we drove around the streets in an open jeep, where anyone could be the enemy and many probably were. Negotiating streets crowded with overladen bicycles, ox carts and mopeds was slow going. By the time we arrived at the air commodore's office on the far side of Saigon near Cholon, four hours had passed since I had left the target area near Rach Kien. I then waited 30 minutes, as the air commodore was busy. He welcomed me cordially saying how impressed he had been watching an O-1 directing air strikes on Cholon only to learn that an Australian had been flying the aircraft. He then enquired how much battle damage the aircraft had sustained. My time in his office amounted to no more than five minutes and he ushered me to the outer office for his secretary to arrange transport back to Tan Son Nhut. It took over an hour before I was advised that transport was ready and I commenced the arduous journey back to the airport. After refuelling '981 I then spent 45 minutes waiting for a take-off slot. Eventually I took off on the taxiway without clearance. Tan Son Nhut had two runways. One used for take-off, the other for landing. It was only the taxiway on the landing runway that was being used.

Back at Tan An the others were waiting for me, half-prepared for a celebration as they all felt bad about the awards ceremony on 20 May. When I told Major Walker what had transpired he was livid and immediately sat down and wrote a report to Colonel Patrick at 19 TASS, complaining about the conduct of my superiors. Although Colonel Patrick was in tune with what was going on and would have followed up Bill's report, I am sure it all fell upon deaf Australian ears.

At 5 a.m. the next morning I had a VR mission but found nothing of note. However, despite the early hour it was always nice to fly in the cool smooth air of the morning. I was back at Tan An by 7:30 a.m., only to be told that I was now on ALO duties. This didn't bother me until I was tasked to go out with Colonel Benson, (call-sign—'Samson 6'), on the 'rounds'. We flew all over the AO to various FSBs, never getting above 500 feet (150 metres) with a 20-year-old, 30-day wonder at the controls of the helicopter. It was very frightening! It was 11 p.m. before we finally finished the day and I was aching all over from continuously clenching my muscles.

On 5 June I had one pre-planned air strike near the junction of the Vam Co Dong and Vam Co Tay rivers, 16 kilometres east of Tan An. There was a lot of evidence of VC presence so we kept probing around with air strikes. The density of local population in the area prevented concentrated B52 bombing. When the smoke settled, I could see we had destroyed four structures among the nipa palm.

The next day I attended the main briefing at Army HQ at 6 a.m. and learned that there was a suspected enemy force just south of The Testicles and the intelligence officer suggested I reconnoitre the area and put in an air strike on the best target I could find. With the fighters due to rendezvous at 10 a.m., I allowed myself plenty of time to really survey the general area of concern. I circled the area around XS 687 547 for an hour at 4000 feet (1200 metres) first and then down low. I could find no evidence of the enemy nor did I entice anyone to shoot at me. As 10 a.m. approached I was desperate to find a target and eventually settled upon a very large healthy looking nipa palm grove with small streams running through it. It occurred to me that this would be the most likely place for any VC to be hiding out.

'Tamale 35, this is Magpie 21.'

Oh hell! I thought. I did not really enjoy working with the Australian Canberra. I guess my baptism of fire was all with high-speed, high-intensity action and the Canberra's bombing procedure was just too slow for me. This attitude was entirely my fault as the Canberras did good work.

The target area I had selected was within a nipa palm grove, probably 20 square kilometres in area. I kept putting my marker rockets into various areas of it but the strong cross winds seemed to have the Canberra foxed as the bombs landed anywhere but where I wanted them. It didn't really matter as we were only probing around a suspected VC area, however, I did find it frustrating. On the Canberra's last pass I put a mark in at the junction of two small streams near the edge of the palms. The speck on the horizon grew steadily larger as the Canberra did its long bomb run in and I gave them a couple of verbal small corrections as the wind had spread my marker smoke horizontally. The aircraft released its bombs and I watched as the bombs sailed right past the base of my smoke, across a couple of rice paddies and impacted 500 metres beyond my mark. The bombs exploded in the bank of a dyke surrounding an abandoned farmhouse and knocked over a few coconut palms.

Just as I was about to clear Magpie off target and advise them that I could not really give him a BDA, out of the smoke ran about 15 black pyjama-clad forms.

'Hey, Magpie, you have stirred up a hornet's nest. I will send you a BDA when available. You are cleared off target.'

'See ya layta, Tamale,' came the reply.

I quickly called Tamale Control and found that the nearest artillery was at Binh Phuoc and I passed the coordinates to the FSB, requesting VT fusing.

'Do you want a marker first, Tamale?'

'Negative, negative. I have troops on the run in the open. Six rounds, fire for effect!'

The sound of VT fusing exploding was different to normal artillery as the rounds go off at about 200 feet (60 metres) above the surface and spray a large ground area with high velocity shrapnel. Nasty, but effective! As the VC were running in different directions, I was going cross-eyed trying to keep them all in sight. The main group were heading for some nearby village huts while two ran toward a solitary large bush in the middle of a rice paddy. Another group of three split off and ran south across the disused rice paddies, heading for a nipa palm area a few hundred metres away. They were outside the splay of the VT fusing impact so I grabbed my AR15 with my left hand and flew down to ground level in chase. As I approached from their rear I opened fire on automatic out of the side window. This was like 'firing from the hip' as I could not aim but, as I banked over the scurrying forms, from 20 feet (6 metres) I could not miss. I saw one go down.

By the time I had pulled up, done a wing over and reversed my direction of flight, the remaining two were in a kneeling position firing at me. This was now a duel.

I started kicking the rudders and weaving to become less of an easy target for them. The muzzle flashes suggested to me they were firing straight at me and I was beginning to regret my impulse in chasing the fleeing enemy. I resisted the temptation to fire early. As I approached the two VC I slammed on 90 degrees of bank and opened fire, emptying my magazine out the left side window at my opponents. As I pulled up into a steep climb again I noted three who had not been cut down by the VT fusing entering a hut, and the other two disappearing into a solitary bush. By holding the control column between my knees I was able to bank the aircraft by moving my legs and at the same time slamming a fresh magazine into my AR15. My intention was to attack my two opponents again but the FSB came on the frequency advising me they had completed their firing. It seemed strange to me that the VC would hide inside a grass hut or under a bush. I decided to concentrate my artillery on the grass hut while I kept an eye on the area around the bush.

To hit an isolated hut with artillery is a difficult call as the charges vary slightly from shell to shell and it is impossible to put two shells in the same place. There is always a small spread. I had the FSB put in a couple of markers. The second one was directly over the hut I wanted to hit so I had ten rounds fired, hoping to hit the target. None of the shells actually hit the hut but some landed only five to ten metres away from it. My aim was to destroy the hut but the FSB were reluctant to fire another round on such a target. When I described the closeness of the fall of shot, the FSB said the concussion of the shells would have rendered the occupants ineffective and firing more rounds would be a waste.

I then flew back over the site of my first encounter and saw the three VC still prone in what appeared to be terminal poses. There was no further ground fire directed at me as I flew around at low level, so I flew over the solitary bush, emptying a magazine into it. I then reloaded and emptied a magazine into the grass hut, which was starting to smoulder from the artillery fire. Further attention to this target by me seemed pointless so I climbed to 4000 feet (1200 metres), well away from the scene and out of earshot, and observed the area through my binoculars. After 30 minutes the hut had burnt to the ground so I started heading back to Tan An feeling quite hyped up but rather pleased with the events. After all, I was still alive!

About a week later, at an early morning Army briefing, I learned a few interesting facts about the day. The two VC who disappeared under the solitary bush later became *chieu hois* (defectors) and told the US Army that the solitary bush actually concealed a small tunnel entrance where they were able to gain refuge from the onslaught of the artillery and my attacks. Prior to that, a group of VC had been sitting inside the entrance to the main tunnel system, watching me probe around the area with the bombs from Magpie 21. The VC were laughing and joking about how we were not getting anywhere near them. My last marker went down 300 metres from them. This brought more laughter from the group, but suddenly an isolated bomb slammed into their tunnel complex. This was Magpie's final bomb—the one that had had me feeling rather dejected. According to the *chieu hois*, there were about 400 buried in the tunnel complex and only about 20 at the entrance were able to dig their way out. Further due to the war, the locals had taken to building heavily fortified log and mud bunkers inside their traditional grass huts for refuge. I had seen many of these mud structures and had wondered what they were. The mud structure usually remained intact after the grass hut was burned off but as I did not see anyone run from the burning hut, it must be assumed the occupants were immobilised by the artillery and perished in the heat of the flames.

Over the next three days I controlled only four air strikes on pre-planned targets, destroying nine bunkers and two structures with no sign of the enemy. On 8 June I did three VR missions for a total of five hours in the air. I did notice that there were telltale signs of human activity on Tan An Island, which was 11 kilometres northeast of Tan An, up the Vam Co Tay River. I reported this to the Army but they said I must be mistaken as there were no reports of VC in the area and the FSB used Tan An Island every morning to zero its artillery guns. The island was about three kilometres in diametre and the dead centre of it was heavily pockmarked by craters caused by the artillery shelling. Tan An Island was formed by a large loop in the Vam Co Tay and, to save navigating the length of the loop, the industrious locals had dug a canal across the neck, forming the island. This canal was named the Ngang Canal and it was less than one kilometre long. Following the river route meant a journey of more than 12 kilometres.

As the island was deserted with no signs of farming activity, there should not have been anyone there. Further, it was a no go zone due to the daily artillery. However, the signs of activity I observed were around the perimeter of the island, in the nipa palm groves along the banks of the river. I could see well-worn paths around the tree lines, discoloured vegetation, wisps of smoke filtering up through the palms and there were mud streams in the river. The VC were probably getting quite cocky as they had not been detected for a long time I would say. They might have felt secure hiding in a no-go zone. Directly across from the island was a canal, the Kinh Bac Dong, connecting directly with the Bo Bo canal and Cambodia, which is the reason I searched this area in the first place. I requested air strikes but the Army did not approve it.

There was always this rivalry between Army and Air Force, which frustrated me many times. The Army would not listen to the Air Force unless they themselves had intelligence information to back it up. After wasting so much ordnance on bombing footprints where the VC had been, it was particularly frustrating not to get air strikes approved on a ripe target. For example, Army intelligence claimed there was evidence of 500 VC in tunnel complexes between Rach Kien and the Bowling Green, 14 klicks northeast of Tan An. I had been over this area many times and could see no such evidence. I was allocated three sets of fighters on 10 June, two sets of A37s and one pair of F100s. The first A37s were carrying an extraordinary mixed load of 750-pound slicks, napalm, RP and 20 mm cannon. If there were VC there, they did not respond and we received no ground fire. After the fighters had left, I flew low around the area for 45 minutes but could only make out a few bunkers and two structures we had destroyed. Perhaps we buried the VC in their tunnels, perhaps not. As far as I know, there was never an insertion of troops to thoroughly reconnoitre the area. It annoyed me to be wasting ordnance on an intelligence report when I knew there was something worthwhile attacking back at Tan An Island.

About this time Saigon and, more particularly, Tan Son Nhut Airport started receiving nightly salvos of 122 mm rockets fired by the enemy from rubber plantations west of the city. We were tasked to orbit this area from sun down to sun up every day. This was in addition to our normal daily tasks of pre-planned air strikes and VR missions, so we were working around the clock again. This was good for clocking up hours in the flying logbook but bad for my rest and health. The rocket watches went on for a week, at the end of which I was feeling jaded after doing two three-hour missions every night. On the watch we would do race-track patterns at 5000 feet (1500 metres) to the west of Saigon, looking out for any telltale flashes of rockets launched or impacting. The rockets were fired on timers so, even if we had seen them launched, the enemy would have been long gone by the time we could get ordnance on the site. Personally, I never saw any rockets launched and the biggest challenge I found was keeping awake while flying around in the smooth air listening to the dedicated rocket-watch frequency, which was mostly silent. I tuned into the busy Tan Son Nhut control tower frequency on another radio just to keep my mind occupied.

A change of pace occurred on the night of 16 June when I was scrambled in the early evening. A VC battalion had been sighted only two kilometres north of the Tan An airstrip. By the time I got there I could not see, using the night binoculars, any sources of heat indicating enemy presence. Artillery were keen to have a fire mission so I lit up the area with flares for 20 minutes in case I could make out any visual signs of the VC. I then pulverised the nipa palm-clad bank of the Vam Co Tay River, as this was the most likely place for the enemy to be hiding. Flying around at 200 feet (60 metres) in the light of the flares and then in the dark did not tempt the enemy to open fire on me so I returned to Tan An and my infrequently occupied bunk. As I lay there waiting for sleep to overtake me I thought this last location significant as it was between Tan An and Tan An Island, where I had seen VC activity on 8 June. Just after midnight I was scrambled again to another sighting in the middle of the Bowling Green, eight kilometres east of Tan An. The same events transpired as had earlier in the evening, with no enemy contact. I was back on my bunk by 3 a.m., totally unrelaxed. I felt there was something brewing.

On 17 June I was given two pre-planned air strikes just north of Tan An Island. Intelligence was *now* suggesting there was VC activity, which they had said did not exist when I had previously reported it. The bombing revealed, and destroyed, four structures, three bunkers and five sampans. When I landed back at Tan An I went straight to the intelligence section, reported what I had seen and got my clearance for air strikes on the island. The intelligence officer told me that they had reports of VC massing in the general area to attack Tan An Army Base. The attack was planned to occur on the anniversary of when they last had overrun the base in 1967.

The next morning I was determined to find the enemy. This was shaping up to be more like what I had been doing in May. I asked Jake, my crew chief, to load extra hand-held smoke canisters behind my seat and received the usual, 'Shit man!' from him. The Army gave me the coordinates of XS 478 724, which placed the approved target right in the middle of the island, where the artillery impacted every day. Before the fighters arrived I tried unsuccessfully to have the coordinates moved but we were permitted to move the pre-planned strike by one kilometre only. I had to use some imagination to fit the target I wanted to hit into the clearance area. I flew at 4000 feet (1200 metres) outside the perimeter of the island and surveyed it through my binoculars. The VC were being even more blase than they had been on 8 June. Several times I saw VC soldiers moving alongside the nipa palm from one camp to another. All the activity was in pockets along the riverbank with nothing happening more than 100 metres from the banks of the Vam Co Tay River.

With my first fighters overhead I rolled in for my marker run on an area where I had noticed VC movement. My rocket caught the enemy unaware and they panicked when the white phosphorus smoke billowed up among the palms. About ten VC ran into the open. I had just enough time to give the attacking fighter a correction to my marker. The 500-pound bombs fell among the fleeing VC, enveloping them in a deadly cloud. When the smoke cleared they were no more. The rest of the first fighter's ordnance, which included CBU and 20 mm cannons, I directed through the nipa palm.

As I finished with one set of fighters, another would check in. Most of them were simultaneously arriving so only the minimum of briefing was necessary each time, as they had witnessed some of the attacks of the fighters ahead of them. It went like clockwork with very little return ground fire. The enemy had been caught by surprise and not near their weapons. As a group of VC scurried across the open fields one second, they were nonexistent the next after the ordnance had run through them. It was impossible for them not to have been killed. With the VC we caught in the open I was able to confirm 28 enemy soldiers killed in various sized groups.

While the fighter attacks went like clockwork, it still took me three hours and ten minutes to direct three pairs of F100s, one pair of F4s and one pair of A37s. At the end of this sortie I recovered to Tan Son Nhut in the early evening. There I refuelled, rearmed and went on a three-hour rocket-watch patrol that was, as usual, non-productive. After I had been on station for two hours I was contacted by Tamale Control and advised that Tan An Army Base was under ground attack by a VC force of unknown size. As it happened, my work on Tan An Island had hastened the planned enemy attack on the base, but with diminished forces. When I approached Tan An from the north the tracer and mortar fire, originating and terminating inside the base, was visible from several miles out. The ground forces were handling the attack well and TACAIR could not be used as the enemy forces were too close to the wire. I circled overhead in case TACAIR was needed and kept the base lit up with parachute flares that were being delivered by 'Spooky', an AC47 flare ship. When Spooky ran out of flares I used artillery flares from the FSB. Merely keeping the area lit up like day prevented the VC from mounting an all out attack on the base. A subsequent US DFC recommendation for me for this day said that my onslaught on Tan An Island was credited with thwarting the effective enemy attack on the Tan An Army Base.

The gunfire slowly subsided and, with my fuel gauges reading near zero, I landed back at the Tan An airstrip. An armed convoy of three M-132 APCs waited to escort the jeep taking me back to the base for safety. Two APCs were ahead of my jeep, closely followed by the third APC. We drove at breakneck speed with the headlights off. Jake was a passenger in my jeep saying, 'Shit man!' all the way. The township seemed totally deserted and not one light shone in any building. It was an eerie journey. The base remained on full alert with perimeter guards armed with spyscopes scanning the immediate area outside the wire. I went to bed exhausted.

I had been in bed a couple of hours when a series of deafening explosions converted my sensuous dreams into terrifying reality. The siren was wailing and men were shouting. I grabbed my AR15 and threw my flakjacket and .38 on over my army issue undershorts. Quite a sight, I imagine, but no-one stopped to comment. With combat boots unlaced, I stumbled along with the multitude in a race to the nearest bunker. I struggled to shake off the effects of my deep sleep. As the incoming rounds of artillery pounded continuous rattling of automatic weapons fire commenced. The bunker I was in was close to the southeastern perimeter wire and looking out through the small viewing slits I was alarmed to see dozens of black pyjama-clad VC running towards us. The artillery flares were lighting them up as the .50 calibre machine guns racked them. I was in awe of the sheer guts and determination of these enemy soldiers. This was a new dimension to me, being on the ground during an attack, and I was terrified. The scene was surreal—just like I had seen in war movies. I had difficulty comprehending the fact that I was a part of it. With my AR15 set to automatic I fired at the illuminated human forms. Some were by now trying to climb the perimeter wire and I could clearly see their faces from 20 metres away. There were enemy soldiers dead and tangled in the barbed wire perimeter and many lay dead at the base of the fence. With about ten US soldiers firing through the slits in the small bunker I was in, the noise was deafening. Occasionally an enemy bullet would find its way through one of those small slits and one of our soldiers fell backwards, hit by an unlucky round.

This attack was over in ten minutes but it felt like an eternity. I became conscience of the VC starting to turn and run away from us, melting back into the night. Then there was the eerie silence, which I could sense rather than hear because I had a loud ringing in my ears that lasted for a couple of hours. In the distance we could hear Vietnamese voices screaming and shouting. None of us left the bunker and we slept with our guns between our knees, squatting against the wall of the bunker. At daybreak we returned to our huts to get some rest. This luxury was not afforded to the grunts as they had clearing patrols and cleaning up to do. Before going to bed I went across to the wire to look at the dead VC. There were dozens of them. One thing I noticed was that most of them were very young. Later I learned from the intelligence officer that most of the attackers were teenagers who had been liquored up on rice wine by the 'real' VC, given a gun and told to go and kill the GIs. What a stupid war!

Intelligence confirmed that the attack on Tan An had been initiated from the area I signalled out on Tan An Island. Around 8 a.m. I was airborne again as the other FACs had been out all night. I put in two air strikes at XS 473 740 just north of the island as directed by intelligence. Although I found and destroyed nine bunkers and two structures, there was no indication of enemy soldiers. I used my wiles again and squeezed targets on the island into my approved coordinates. Once again, my first marker rocket produced what appeared to be a swarm of ants as the enemy poured out of the nipa palm. They were probably resting up after their all night attack on Tan An. The VC on the island had become so overconfident about their security that they were not dug-in and were camping on the surface. As there were no friendly troops or civilians anywhere near I cleared the fighters for random heading attacks. Before the VC could disperse, I fired a Willy Pete among them and they became enveloped by the thick white smoke. I was able to tell the fighter, 'Hit my smoke!' Due to the random headings the fighter was releasing a pair of 500-pound bombs before any of the VC appeared outside the smoke of my marker. The exploding bombs extinguished my smoke. None of the ten or so VC could have survived the explosion.

I could see panic and disorder within the VC ranks as they scurried about like mice rather than fire their weapons at us. By using random headings, the VC did not know from which direction the next fighter was coming or when. The usual race-track patterns would have inevitably caused the fighters to not only attack from the same direction, but also to attack at set intervals. If the attacks were from, say, the north, all the VC had to do was set up an umbrella of automatic weapons fire to the north at regular intervals and the fighters would fly into it. There would have been no tracking required on the part of the enemy. As our attacks were using random headings the fighters were over the target at irregular intervals. One fighter might attack from the east and one minute later, another from the south. Next, one would come in from the west within 20 seconds and the next might be in from the north two minutes later. It must have been very confusing for the VC. And, it was very demanding on the FAC. On this occasion I had three F100s in the formation, which kept me busy staying out of the way, keeping the fighters in sight, as well as monitoring the VC locations. I might be heading west and a fighter calls in from the east requiring me to steep turn through 180 degrees in a nose-high wing-over manoeuvre to keep him in sight. It was like a mix of low-level aerobatics and tag-team! After a short time I was into a rhythm and on most of the fighter attacks had myself facing him as he rolled in.

Visual contact with the enemy was almost continuous but there was very little return ground fire from them. On several passes the fighter pilots could see the enemy in the open themselves and needed no direction from me. The excitement in the fighter pilots' voices showed this was an uncommon event for them. When the ordnance got down to 20 mm cannon, rather than clear each pass, I allowed the fighters to pick their own targets. I assisted by calling when I was circling directly above the VC. It was basically a turkey shoot. On one pass I could see six VC running across an abandoned rice paddy from one nipa-palm clump to another. The F100 was tearing up the ground behind them with 20 mm cannon fire. The pilot then walked his fire up through them, virtually picking them all up and blasting a large hole in the nipa palm. If one could consider that death—in some macabre way—could be beautiful, that pass was beautiful. The F100 pilots were so exuberant about the engagement that, as they left, they did a low pass and climbed steeply away doing several rolls.

A couple of days later I was having a few drinks at the fighter pilots' hooch at Bien Hoa. One of the pilots who had been on the mission to Tan An Island showed us his gun-camera film. On one pass the film clearly showing one very brave VC standing in the open. He was facing the oncoming camera, which was attached to six tons of death-delivering machinery travelling at 450 knots (730 kilometres per hour), and firing his AK47 at the screaming monster. The ultimate defiance. It left a knot in my throat. The VC just disappeared in an eruption of dirt and debris.

While I was in the middle of this grim carnival, a VNAF Huey turned up looking for some action. I do not know if he was shot down or went all thumbs and elbows, but at some stage he crashed into the river. I did not see it happen but there was nothing we could do so we continued with our attacks. During one of my orbits I could see the evidence in the water of where it had crashed but could not see any bodies or survivors. During the attacks I kept asking Tamale Control to get the Army inserted but it did not happen. As Tan An Island was just that, it would have required a mobile riverine force or airborne force to be inserted on the island, which would have taken time even if they were available. To me, fighting my own little war out there, I could not appreciate the lack of Army support at the time.

A Huey gunship was dispatched, which I was able to direct onto a couple of locations and the crew gave it their full attention with rockets and .50 calibre guns. I did see three VC enter a small canal and disappear under the water. I had my own AR15 cocked and ready waiting for them to surface but after ten minutes I drew the conclusion that they were all Johnny Weissmullers or they had entered a tunnel complex through an underwater entrance. The gunship killed a further four VC on top of the 14 I could confirm. I was very disappointed after all my efforts over the last three days around Tan An Island that the Army did not do a sweep through the area. I had to accept that they knew the overall picture, while I was only a small cog in the massive machine. Although I could only confirm 48 enemy killed in the open, I am sure a ground sweep would have shown a much higher tally.

At 5 a.m. the next morning I was scrambled to assist troops-in-contact at XS 864 658, seven klicks southeast of Can Giuoc. Very accurate bombing by two F100s and two F4s had the grunts up and moving again with me spending only one hour and 40 minutes in the air. There was no return ground fire and I wished they were all that easy!

The next two days were spent on seven pre-planned air strikes and doing rocket watch over Saigon. I was also able to call into Bien Hoa and have '981 serviced—as well as myself at the fighter pilots' hooch. After a nice pizza and a few rum and cokes I headed back to Tan An. When I arrived back there was a celebration taking place. We were moving base to Dong Tam. Damn, I thought. This meant not doing the Saigon rocket watch any more! My possessions amounted to three flight suits, two pair of boots, one hat, six pairs of underpants and socks plus my guns. Packing was not going to be an issue.

SOCK IT TO 'EM BABY

On 23 June I flew '981 to our new home at Dong Tam, which was deeper into the Mekong Delta, 25 kilometres southwest of Tan An. The base was on the north bank of the My Tho River, which was the northern tributary of the mighty Mekong River. It was situated eight klicks due west of the town of My Tho. *Dong Tam* means, 'united hearts and minds' in Vietnamese, and was the name personally selected by General Westmoreland. The construction of Dong Tam was started in early 1966 and was created by dredging sand from the bottom of the My Tho River and filling in 640 acres (260 hectares) of inundated rice paddies. Some rice paddies were dug out to form a basin for river craft of the MRF (Mobile Riverine Force). The island formed by the dredging was a dust bowl in the dry season and a quagmire during the wet.

The airstrip was built from Marston matting and it ran parallel to the Xang Canal, which was at right angles to the My Tho River. Marston mats were used during the Second World War to build runways in jungles and during the Korean War on boggy ground. Where possible we would take off into the north and land into the south due to the unlit high masts on the river craft off the southern end of the runway. Unfortunately, the FSB was situated 200 metres off the northern end. The artillery caused us considerable concern while taking off and landing. Along one side of the airstrip was the helicopter operating strip. The helicopters would sometimes produce uncontrollably turbulent air across our landing path.

While the airstrip left a lot to be desired compared to Tan An, it was nice being on a relatively secure base where one could go for a walk without wearing a flak vest and guns. There was also a good base exchange for shopping and the Red Cross girls were also based at Dong Tam. We would sit outside our hooch waiting for one to walk by and we would survey them out the corner of our eye through dark glasses until they disappeared from sight. Some would say we were perving but under the circumstances I would rather think we were admiring the female form. Despite the general improvement, the ablutions still left a lot to be desired. Showers were still rather public but the latrine was three holes, for the lucky officers—other ranks had a ten holes.

Besides the 3rd Brigade and 9th Infantry Division Headquarters, there were numerous other units based at Dong Tam. One of the most notable was the Mobile Riverine Force, which was attached to the 2nd Brigade. The MRF conducted their operations from US Navy armoured troop carrying boats. The boats were preceded along the waterways by minesweeping craft and the whole flotilla were escorted by armoured gunboats called 'Monitors'. The Monitors were armed with 81 mm mortars, 105 mm howitzers and 40 mm machine guns. Many times this armada would be ambushed from the heavily foliated shoreline by enemy using rockets and recoilless rifles. By the end of 1968, areas in the Mekong Delta, which were previously held by the enemy, were more accessible by friendly forces thanks to the MRF operations.

We spent 24 May setting up our new abode and I flew up to Tan An to collect my crew chief, 'Shit, man' Jake, and a few other comforts we did not want to leave behind. I had no combat duties for a couple of days and tending to homely duties made me feel almost human again. The next day we were back to our normal routine. Instead of going for a sightseeing flight around our new AO to get to know the area we launched straight into pre-planned air strikes. I had been borrowing money from the other FACs because I had not been paid since arriving in Vietnam. As my second and third air strikes were in the direction of Vung Tau, Major Walker suggested I put in the air strike and then refuel at the RAAF base instead of back at Dong Tam. While at Vung Tau I could receive my pay and then get back to put in the third air strike.

Nothing out of the ordinary was experienced during the first two missions except that it was obvious the VC were not there and we were bombing footprints and unoccupied bunkers. After I parked my aircraft in the USAF lines at Vung Tau, I found my way across to the RAAF HQ area. There was no-one there and the pay section was closed. I then went to a hanger where RAAF mechanics were working on a RAAF Huey and pronounced, more as a statement than a question, 'The pay section is closed?'

A RAAF sergeant replied, 'Of course, it's a holiday!' Being a weekday in a war zone, I could only think to say, 'I beg your pardon?' I cannot remember exactly what answer I received but the sergeant laughed and told me that HQ staff were down at Vung Tau Beach waterskiing. Only the 'sharp-end' were working. I explained that I was an Australian with the US Army and had not been paid since arriving in country and that I had a valuable aircraft lying idle. I needed to get paid and get back to work. The kindly sergeant gave me directions to the beach and offered me his jeep to chase after the pay officer. As I drove off the sergeant called after me, 'Best of luck—you could be lucky, Sir!' I wasn't sure if he was wishing me well or indicating the fruitlessness of my task.

When I arrived at the beach I was not prepared for what I saw. Sure enough, they were waterskiing but all the men were wearing bathing suits and some were in the company of bikini-clad Vietnamese girls. I spotted the pay officer loading up his plate with steak and onions at the barbecue. He greeted me cordially but when I told him I was there to get paid, his attitude changed somewhat. He told me it was his day off and I would have to come back another day. After I advised him that I was trying to fight a war despite him, he told me that that was my problem, and things got ugly. Before I knew what had happened, I had the pay officer by the throat with his steak and onions hurtling across the sand. Someone pulled me off my target and Group Captain John Hubble appeared asking what the trouble was. The pay officer got in first by saying I had demanded to be paid on a holiday and had attacked him. I had dropped my AR15 in the sand and was trembling with rage. Group Captain Hubble took a few seconds to calmly assess the situation and spoke to the pay officer saying, 'This man is a combat pilot and has obviously just come from a shooting match somewhere. You will pay him now.'

He then turned to me and said, 'Don't let this worry you, I will sort it out. Take care', and he went back to his barbeque. On the drive to pay section I tried being civil to the pay officer but he was carrying a grudge. After I collected my pay, I thanked him and drove the jeep back to the maintenance hanger, leaving him to find his own way back to the beach.

Group Captain Hubble had been through the Second World War, the Korean War and the Malayan Confrontation. Vietnam was his fourth war. A lot of junior officers did not have much of an opinion of him. The knockers amplified anything uncomplimentary he had done, but the things he had achieved were not discussed. I have to admit that I was also guilty of ridiculing him without foundation, only because other people did. It unfortunately seems to be the Australian way. After this incident I was not heard rubbishing Group Captain Hubble ever again. The thing that got to me most about this whole situation was that the pay officer received exactly the same war benefits and campaign medals as myself. Our daily lives in Vietnam were worlds apart. With the Australian system, a person who has been in combat cannot be differentiated from those who have done the logistical support tasks and not seen combat. For every combatant there are about six in logistical and support roles.

Due to the delay I experienced at Vung Tau, I had to fly at full throttle to XS 374 555, 13 klicks northwest of Dong Tam, to make my rendezvous with my third set of fighters. As it transpired I only had a couple of minutes to look over the target area before the two A37s checked in. This area centred on Cai Lay, which was so infested with VC activity that it was almost impossible to drop a bomb and not hit something the enemy had constructed. With 750-pound bombs, 250-pound bombs plus RP we were able to destroy one structure and three bunkers.

After a long and frustrating day, I was looking forward to a shower and bed. On arrival at the hooch, Major Walker asked out of courtesy if I had been paid—I did owe him 20-odd dollars. When I told him what had happened he said, 'Y'all seem to have one fucked up organisation'. This, on top of the awards ceremony debacle, must have been a bit much for him to comprehend. I was too tired to disagree—he was getting a flood of bad impressions and there were some good aspects to Australians he had not yet seen.

I headed for the showers, doing my usual trick of removing my boots only and walking into the shower without undressing. With the cold water running on me, I would slowly undress, washing each item of clothing in turn. These I would then put on clothes hangers and hang them on my ceiling fan to spin dry.

8

ENGINE FAILURE

AFTER MY CONFRONTATION WITH the pay officer at Vung Tau, the last thing I wanted to do the next day was an ALO duty, which to me was soul-destroying. At 5 a.m. I was dragged out of bed as the Army were in contact with VC troops near Cai Be, 28 kilometres west of Dong Tam. It seemed like only yesterday that I had done my last ALO duties and I was sure it was not my turn again. However, I did not argue, as I was becoming very robot-like in my actions. I joined the CO of the unit involved, a lieutenant colonel, at the chopper pad, from where we took off in an OH23 'Raven' with a fearless 20-year-old warrant officer at the controls. His name was George C. Gray from Melrose Park, Montgomery County. This was his first flight back after R and R in Honolulu, where he had been married to his high-school girlfriend, Kathleen Vespico. We cruised along at 200 feet (60 metres) with my knuckles turning white due to the death grip I had on my seat. It was pleasant slipping along in the smooth morning air but I kept imagining one of the VC stepping out from behind a tree and hosing us down with his AK47. At that height we were a plumb target. Army aviators seemed to put personal safety last. While we took chances as Air Force pilots, it was only when necessary.

The colonel was talking with his troops well before we arrived over the contact area so we had a good idea of the lay of the battle when we arrived. The colonel decided not to use TACAIR and directed artillery himself onto the target area. This made me somewhat redundant and I was wishing I were not there. I hunched up into as small a target as possible and clenched my torso tight. This only helped psychologically but it was not the sort of situation where I could just sit back and relax. For about 30 minutes we tore around at zero feet, twisting and turning. It is a good thing I did not suffer from airsickness. Most of the time, I could hear gunfire through the open sides of the Raven. My ear had now tuned into the difference between M16 and AK47 gunfire. Both sounds were evenly interspersed. Fortunately I could not hear any heavy calibre enemy gunfire. The gunfire was punctuated with artillery explosions. Although terrifying, it was interesting sitting there as an observer as I noticed the AK47 gunfire progressively diminish as the artillery took effect and the tide of the battle changed in favour of the US ground unit. On three separate occasions I watched in horror as small holes appeared in the bubble glass in front of us. Each hit would have me mentally reduce my body size and flinch in anticipation of the next hit, which did not come immediately as I expected it would.

Once the area was secure the colonel had WO Gray land in the rice paddies while he briefed his platoon leader. It was a great relief to me when we took off again and headed back to Dong Tam. When I arrived back at the hooch I was greeted with the sight of two very pretty Vietnamese girls doing chores around our living quarters. As we were officers, the Army provided them to keep our hooch neat and tidy. They were both very sweet, didn't speak English well but smiled anytime you looked their way. When they noted my showering ritual they laughed and soon had me trained to give them my soiled clothes before I entered the shower. It was a little embarrassing at first to have them stand there while I undressed but I soon got used to it. They seemed to be present from first light in the morning till well after dark, seven days a week. Their names were Mary-Ann, who was married, and

Sophie who was single. These would not have been their Vietnamese names but Mary-Ann and Sophie were names we could get our Western tongues around. I do not remember exactly what their pay was but I thought at the time it was rather dismal, considering the hours they put in.

I referred to the FAC Hooch as 'our living quarters', but that's a little misleading. Living quarters conjures up an impression of something like a home. At Dong Tam the quarters were the best I had seen on any Army base but nowhere near as comfortable as on Air Force bases. We had a slightly improved version of what I had lived in earlier. In the little spare time I had available I planted grass down one side of the hooch, not only to get a little greenery into an otherwise desert landscape, but also to suppress the endless dust that blew through the open sides of our dwelling. At one end we created a small air-conditioned crew room, where we could sometimes escape the heat and 100 per cent humidity. This room was built from plywood and other materials we had stolen from various locations around the base.

Right next to our hooch was the quarters of the 9th Infantry Division band. They had built a sound shell in which to practise their music, but unfortunately it faced towards our hooch. Infrequently the members of the band would practise various instruments solo, which was annoying. Sometimes, however, a group would get together and have a jam session that would include anything from Dixieland to Blues. These sessions were most enjoyable.

My schedule for 28 June proved to be busy, with me controlling seven air strikes. Major Richard Nelson had just joined the 3rd Brigade and he was taking over from Major Walker, who was assigned to HQ 19 TASS. Major Nelson was a quiet, unassuming type with a pleasant manner. I was to provide his combat training so he had clambered into the back seat of '981. The task we were set was a VR mission at WS 945 439, which was 15 kilometres northwest of Vihn Long and 43 kilometres from Dong Tam. This was the furthermost west we would operate with the 3rd Brigade. The area was a free-fire zone, which meant anything there was enemy and prior approval was not required to return fire when fired upon. After orbiting the area for a few minutes I started to pick up unusual signs of activity. Being a free-fire zone there should not have been any civilian traffic and yet there were obvious signs that a number of foot trails had been recently used. Some of these trails led to what seemed to be abandoned huts with vines growing over them as nature tried to reclaim her territory. From the roof of one structure I could see wisps of smoke filtering up through the grass matting. The first pair of F100s arrived overhead and I briefed them from 4000 feet (1200 metres). The occupants of the hut would not have known about the death delivering arsenal above them. As it was a free-fire zone, it was not necessary for me to mark the target and I was able to describe to the pilots the structure I required hit. The first F100 rolled in from 10 000 feet (3000 metres) and straddled the hooch with 500-pound bombs. Only one VC ran from what was left of the hut and he was quickly disposed of as he ran in the open by the second F100 with 20 mike-mike. The rest of the ordnance we dumped on the same spot and another hut nearby.

Straight after this three F100s checked in and we followed the same procedure. As we no longer had the VC by surprise I did not expect to see any more enemy. However, we caught two VC in the open and they were killed with 20 mike-mike. I was allowing the fighters to use random headings, which they obviously enjoyed doing and this method kept constant pressure on the target area. At no stage did I detect any return ground fire and I became concerned that I might be attacking civilians. However, on some of my lower passes I was satisfied that they were all young men and that there were no women or children present. Attacks by the third set of F100s pushed six VC out into the open. They were running towards a hut where I suspect there may have been a bunker system. An F100 neatly placed two canisters of napalm behind them. One canister hit first and engulfed the fleeing forms. The second canister virtually went through the door of the hut. For an instant I could see the napalm spread out through the back of it before the whole scene became one giant, glowing inferno.

ENGINE FAILURE

During the fourth fighter attacks I heard and felt what I thought was an RPG round explode beneath us. The source of the RPG was obvious and we covered the area with 750-pound bombs and 20 mike-mike. As the fifth fighters were attacking, I heard a series of alarming explosions that I initially thought were RPGs again. I was about to warn the fighters of ground fire when I noticed that an F100, in the late stages of his attack at low level, had artillery shells exploding very close to him. I immediately told the fighters to hold high and north of the target where I knew no artillery could be coming from and called Tamale Control. Sergeant Cover, who listened on UHF to all our air strikes, was right onto it and came back with the location of the FSB. It was an ARVN unit at Sa Dec. This brought a lump to my throat as I had been right in their line of fire and the artillery shells must have been falling through my flight path. It took a few minutes for Sergeant Cover to get the ARVN to stop firing. Once again, the ARVN had seen an attack going on and decided to get involved without approval. In the meantime my fifth set of fighters were running short on fuel and had to leave so I had them do one last pass to salvo their remaining ordnance on the target area.

Although I had covered an area of about two square kilometres, there were a couple of more targets I considered needed attention. There were no further fighters available within my endurance time but Sergeant Cover located a pair of Huey gunships looking for business. I briefed them on where I thought some VC may be and then sat back and watched them in awe as they went about delivering RP and .50-calibre gunfire. They were a delight to watch and they really gave the area a going over. The gunships even flushed out two more VC, whom they hovered over and blew to pieces, reminding me of a cat and mouse situation.

Back at the base Major Nelson must have been impressed with what he saw as he recommended me for an Air Medal for Valor. He was somewhat surprised when I told him that my government would not let me accept it. The look of disbelief on his face said it all. He looked at Major Walker for reassurance. Bill Walker shrugged and said, 'The Aussies have a unique procedure where they believe, to get the best out of your men, you do not show them any credit. Get Cooper to tell you what they did to their Army at Long Tan sometime.'

On 27 June I got my second day off for the month but I could not relax. The hectic pace of day-after-day flying had me so hyped up I could not unwind. I spent the day jumping, from working on maps, to pottering about the hooch, to wandering about the base. Finally, I asked Major Walker if I could do a VR mission but was told no, to go and relax. Being so wound up I could not concentrate on any one task for much more than a few minutes. Perhaps I had become hyperactive.

On the last day of June I was also not tasked to fly again but I convinced Don Washburn to let me take over his mission. The pre-planned strike was at XS 287 574, 22 kilometres northwest of Dong Tam in the 'VC Panhandle', as we called it. In 1993, when I was flying out of Saigon for Air Vietnam on loan from Ansett Airlines, I viewed some wall charts in the bunkers of the Presidential Palace. On the chart were plotted all the enemy locations through the Panhandle. There were so many tunnel systems there that it is a wonder the whole area did not cave in. After flying around the allocated target area for ten minutes I selected a target which seemed a little spoiled compared with the rest of the area, in that it showed signs of human activity. An Australian Canberra checked in and we spent a leisurely 30 minutes dropping 500-pound bombs on it; the result being four bunkers destroyed.

July started with one pre-planned air strike, which I used to give Major Nelson another 'first' instructional flight. The 28 June lesson was interrupted by the mission becoming so involved—there was not the time for any real instruction and his baptism by fire proved to be rather hectic. Now having just one uncomplicated air strike enabled me to demonstrate copybook procedures. Major Nelson insisted on going in the back seat as he considered he would learn more by watching someone experienced. This was a completely different approach to the one Major Walker took, who just threw me in at the deep end. Out of the blue Dick

ENGINE FAILURE

Nelson said, after the air strike, 'Let's go up to Saigon and sort out your people on this awards nonsense.'

I told him about the efforts Bill Walker and General Ewell had made but he felt confident he could plead to their sense of fair play. On arrival in Saigon we both went across to 7th Air Force HQ, where Dick Nelson borrowed a jeep. I went back to Tan Son Nhut and serviced the aircraft before returning to Dong Tam. I think Major Nelson was planning on a few drinks with buddies in Saigon. And sure enough, just before dawn the following day when I flew up to Saigon to retrieve Major Nelson, my suspicions were confirmed. His haggard appearance showed he had participated in more than just an aperitif. However, despite how he must have felt, he was waiting at dispersal on time and I did not even have to shut down the engine. On the way to Dong Tam he did not say much about his meeting at HQAFV but did say that he would continue to try to get my awards through.

On approach to Tan An, Tamale Control diverted us to troops-in-contact at XS 130 560 in the VC Panhandle. I only put in one pair of F100s and the grunts were then satisfied that the resistance had been broken. After the air strike I climbed up to 4000 feet (1200 metres) to show Dick Nelson some of the AO. Near the junction of the Thap Muoi and Da Bien canals I was surprised to see about 20 VC walking along the bank. Every time I found VC in the open in daylight I always got the uneasy feeling that they might be friendly militia. I checked with Tamale Control, who in turn checked with Saigon and confirmed that they must be enemy. Two pairs of F100s were diverted to work my new target.

I briefed the fighters from high altitude and rolled in for my marker. My Willy Pete hit very close to the VC, who split into two basic groups. As the VC were out in the open, I cleared the F100s in with 20 mike-mike first. I estimated that the fighters killed seven of the enemy in the first pass. I then had them drop their bombs on the target just to get coverage. As the debris settled I saw another ten VC running for a small canal bank. The lead aircraft of the second pair of F100s did a repeat of the previous pair and tore up eight VC with 20 mike-mike. I saw two VC disappear into the reeds so I had napalm and 500-pound bombs spread along the bank. I did several low passes over the area and could see some bodies but no tunnel entrances. Once again, Major Nelson was missing out on his instruction but he was getting plenty of valuable combat demonstration as he quietly sat in the back seat saying nothing. Or so I thought. On the way back to Dong Tam I looked back and there he was, mouth wide open and sound asleep.

My first flight on the 3 July was a VR mission in the Ca Duoc area. On arrival I could see there was some helicopter activity indicating an Army contact. Tamale Control gave me the frequency of the unit, which was the 2nd Battalion, 47th Infantry of the 1st Brigade. I could not see a FAC on station so I called the command chopper.

'Panther 6, this is Tamale 35. Do you read?'

'Tamale 35, this is Panther 6. Go!' It was the voice of Lieutenant Colonel Frederick Van Deusen, a 37-year-old from Fayetteville, North Carolina, who had just taken command of the 2nd/47th three weeks earlier.

'Panther 6, do you require any TACAIR or FAC involvement?'

'Negative, Tamale. But many thanks for asking. We are just mopping up.'

My eyes then picked up the command chopper turning low over the Vam Co River. I glanced away and when I looked back, to my horror, I could only see boiling water and no chopper. He had either been shot down or lost control. Other choppers were quickly on the scene so I reluctantly departed to complete my VR mission. As it transpired, there were four KIA and three survivors from the crash. Van Deusen's body was recovered some time later.

Lieutenant Colonel Van Deusen came from a military family. He had a brother, Edwin, who was also a lieutenant colonel, and their father was Colonel Edward Van Deusen. Fredrick Van Deusen was General Westmoreland's brother-in-law and, ironically, he was killed on the same day that the general was sworn in as the US Army Chief of Staff. Sadly, Fredrick's wife and two sons had all passed away by 1996. I have also been told that the Van Deusens are distantly related to ex-President George Bush's wife.

ENGINE FAILURE

After the VR mission I put in two pre-planned air strikes, destroying six bunkers 15 kilometres west of Dong Tam. Over the next three days I controlled seven air strikes on intelligence targets but received no ground resistance—the enemy were not there. While I was putting in a pre-planned air strike on camouflaged structures on 7 July I noticed some sampan activity under the vegetation along the banks of a small canal near Cai Be. We sunk three sampans before the fighters ran low on fuel and had to leave. Soon after the fighters left I saw six VC in the open not far from where we had been bombing and I called in an artillery mission. As soon as the first artillery shell hit, the VC spread out so I had the FSB put in a few VT fusing shells. Flying low over the area I could see three bodies.

The next morning I went back to the same area and destroyed three more sampans with two F100s. There was some light small arms fire from a number of different areas so I requested artillery again on a small overgrown stream back from the main canal, as I noticed signs there of disturbed vegetation. This fire mission produced a small secondary explosion, indicating we had hit a weapons cache. On 9 July I was covering the insertion of troops by helicopter at XS 096 540 just north of Cai Nuoc. There were a number of Huey gunships rocketing and strafing around the insertion area so I kept out of their way, remaining handy if they wanted some TACAIR. Lieutenant Colonel Hemphill, the 3/39th CO, was on board an OH23 helicopter, along with his FO, Captain James Latham. WO George Gray was piloting. As I watched the OH23 fell out of the air into open rice paddies, having been hit by ground fire. As there were a large number of choppers in the area, my assistance was not required. By the time I arrived back at Dong Tam, Lieutenant Colonel Col Hemphill and Captain Latham had already been extracted but George Gray was dead before the chopper hit the ground. His body was recovered later that day. I was preparing for a night mission when George was brought in. Where once I had not been too shaken by death, it was now starting to have an adverse effect on my emotions and I was appreciating my vulnerability. George Gray's

21st birthday was the day he died and we had been planning a few drinks to celebrate.

I was spared the time of mourning as my pre-planned strikes evolved into a troops-in-contact situation. After George Gray had been brought down, the 3/39th received increasing contact with the enemy into the late evening. By 1 a.m. I had put in three air strikes in support of the ground troops. Major Nelson came along with me on most of these flights but did not want to fly the aircraft, he seemed quite content to sit in the back seat observing. This was a particularly difficult target, as I could not get the traditional flare illumination by Spooky or artillery in time before the fighters ran short of fuel. Mortar flares were limited, so once I had a fire burning on the ground I had to use that as the reference. It was a particularly black night and a number of times I was startled by the dark form of large coconut palms looming very close to the O-1. I had to admire the attacking fighters with their accuracy under such conditions. However, once I got focused on the ground fire and exploding ordnance I had a set height on the altimeter that I would not go below. Again, heading back to Dong Tam I noted Major Nelson asleep in the back seat.

Being an early morning finish, I was not scheduled to fly that day but Major Nelson wanted to visit the commanders of neighbouring FAC units to familiarise himself with the area. This was an excellent idea but I could not understand why Dick Nelson wanted me to fly him everywhere like a taxi driver. I did not really mind as he was a delightful person and I enjoyed his company. If I was not flying him about I would be bored stiff back at Dong Tam. We flew into Bien Hoa, Can Tho and Bien Thuy, arriving back at Dong Tam late in the evening. It started to concern me that Major Nelson might lack confidence and did not want to fly. However, when I eventually had him do a mission, it was as though he had been doing it all his life and I was quite impressed. I never dared ask him why I was doing all the work but I imagine it was his way of learning by observing and perhaps he was being generous by allowing me to do the flying. His attitude was quite refreshing as I was always used to senior people

My mother, younger brother and I walking along Glenelg Esplanade in Adelaide, South Australia, in 1943.

Olivia Newton John. Subject of my lucky charm medallion.

Antarctica 1962. Left to right: Sgt. Tiller, Plt OT Cooper, Sqd. Ldr. Batchelor and Sgt. Richardson. I can't remember the names of the penguins.

75 Squadron Australia's first Mach 2 fighter squadron. Wing Commander Jim Flemming, CO, is third from the left with yours truly on his right.

My flying helmet retrieved from the wreckage after the action. The bullet missed my forehead by 5 mm.

Sgt. Alan Kisling, Charlie Company, 5th/60th, May 1968.

Captain Andy Anderson examining the wreckage of Captain Ike Payne's aircraft.

Some light relief at Tan An, instigated by Bill Walker. Sending up the Australian award system.

An award made by the surviving members of the 5th/60th for 10 May 1968 action.

Sgt. Howard Querry,
KIA 10 May 1968.

Capt. Edmund Scarborough,
KIA 10 May 1968.

W/O George Gray,
KIA 9 July 1968.

Lt. Col Frederick Van Duesen,
KIA 3 July 1968.

Col. Robert Archer, CO, 1st Bde.

Lt. Col. Anthony De Luca, CO, 3/39th.

Maj. William Walker, ALO, 3rd Bde.

Maj. Richard Nelson, ALO, 3rd Bde.

Yours truly with his trusty AR15 Carbine and faithful '981 at Tan An.

Sgt. David Robe attaching his mascot, a bra, to the antenna of his APC amid the chaos in Saigon, May 1968.

Major Bung Ly with his family of seven landing '981 on the USS Midway in April 1975.

The ill fated Raven 20 minutes before being shot down on 18 August 1968. The pilot is attending to necessary function in front of the bubble.

Having a friendly drink with ex-VC generals and colonels at the contact site of 18 August 1968. Circa 1993. The officer third from left claimed to have fired the fatal burst.

At 3 am in the cockpit of a Boeing 747 on my last international flight as Captain. Penny Hanrahan and Ron Jones joined me from Hong Kong to Sydney.

ENGINE FAILURE

grabbing the limelight and the best for themselves. However, Dick, like most of the Americans I flew with, had done all their initial flying training on tricycle undercarriage aircraft and the O-1 was the first tail-wheel aircraft he had flown. My training centred more around his handling a tail-wheel aircraft on the ground than actual combat operations.

On one early morning air strike, at about 1 a.m., I was running short of fuel near Sa Dec on the western extremity of our AO. Dong Tam advised me that it was raining so I headed for Vinh Long, only to be told that they were closed due to the weather. I turned for my next airfield, which was Binh Thuy. It was pitch black and there was not a light on the ground to be seen. It was starting to rain lightly but, as I could not see the ground anyway, I flew at 1000 feet (300 metres) on instruments and a compass heading. As it started to rain heavier I descended to 200 feet (60 metres) and got sight of the Hau Giang River, which I followed northwest, hoping to get a glimpse of Binh Thuy. With my fuel at a critical level I saw an aircraft cross my flight path at right angles, obviously on approach to an airfield. I called Binh Thuy tower and got a clearance to land. I thought the runway surface at Binh Thuy was bitumen but I was touching down on PSP. I made it down and the engine was still running as I felt my way slowly along the taxiways in moderate rain. The tower reception kept fading and Binh Thuy tower kept asking where I was. I shouted at them that I had no reference and did not know. They kept coming back saying, 'Say again!'

Finally I found the control tower and before I could stop myself from abusively saying, 'What's wrong with you, I'm sitting right under the control tower', I looked up through the rain and read the name, Can Tho. I was on an Army airfield five klicks from my intended destination. Little wonder Binh Thuy could not read me! I felt quite foolish. With so little fuel, I would not have made it to Binh Thuy anyway—there seemed to be a lot of air in the tanks when I refuelled.

During my before-dawn departure the next morning I had trouble clearing my spark plugs. I had already taken the runway so

I taxied clear to let another aircraft take off behind me. As that aircraft was climbing through 200 feet, I saw out the corner of my eye tracers arc towards it. Then there was the flash of a fireball which disappeared among the silhouetted palm trees very close to the artillery emplacements. I commenced my take off 30 seconds later and as soon as I lifted off I threw on sixty degrees of bank, climbing steeply over the base to avoid the danger area. The tracer fire could not have been more than 100 metres outside the perimeter wire. As I climbed out the perimeter guards were racking the area with .50 calibre fire but there was no return fire. The VC obviously took one lucky burst and hightailed it out of there.

My next target was at XS 424 558, about four kilometres northwest of Ben Tranh. Using two F100s we destroyed two structures and three bunkers with no ground resistance. As I wasn't far from Tan An, I landed there to rearm and refuel before heading out west of My Phuoc Tay for a VR mission into the VC Panhandle. Flying at 4000 feet (1200 metres) on low power while looking through my binoculars, I picked up four VC crossing a single-pole bridge over a small canal. I decided to use artillery on them and remained at the same altitude while I passed the co-ordinates to the FSB at Cai Lay. As it was the free-fire zone I did not call for a sighting marker and requested five rounds of VT fusing, giving clearance to 'fire for effect'. The VC were midstream when they were hit from nowhere seemingly by complete surprise. After the firing, I went down to zero feet and all four VC were floating dead in the canal.

I had nothing scheduled for the next day but before daybreak, just as I was having a delightfully sensuous dream, the less than sexy voice of Sergeant Cover interrupted, 'Sir, Captain Cooper, Sir. Troops-in-contact.' Looking back, I think this is all he ever used to say to me.

On a morning I thought was going to be a day off, I started a 7 hour 40 minute shift, controlling nine air strikes. My body was definitely crying out for relaxation but even when I had time off, I knew I could not unwind. Within five minutes I was in the FAC jeep, with Sergeant Cover driving through the chill morning air

ENGINE FAILURE

to the airstrip. It had been raining most of the night and the air was damp with a low overcast. Back in Australia, under such conditions, I would head for the meteorological office to enquire about the prospects of the weather clearing. Here, soldiers were in the field dying and it was imperative that I get out there to help. It hardly entered my mind that the conditions might not be flyable.

As I approached the revetment where '981 was kept to protect it from enemy attack, I could see Jake was already warming the engine for me.

'Good morning, Jake!' I shouted above the running engine.

His response was his usual two syllables. He meant no disrespect but it was just Jake's way of acknowledging the inclement weather. After a quick check of the fuel gauges, controls and rockets I taxied onto the runway, calling the artillery to hold their fire. I looked ahead and the mist and light rain prevented me from seeing the far end runway lights. This suggested the visibility was less than 1000 metres. No sooner had I cleared the artillery guns than they started firing again, ever so close. By 500 feet (150 metres) I could not see the ground so I descended to 300 feet (90 metres) and headed west to XS 192 428, just near 'Snoopy's Nose'. As I glided through the morning mist, daylight started to break, the rain thinned and the cloud base started to lift. By the time I reached the target the cloud was 5/8th at 1500 feet (450 metres) with visibility of five kilometres. Things were starting to look better.

As I approached Snoopy's Nose—so called because the river had a large bulge where it looped back on itself, forming a shape which from the air looked like the comic character's nose—I called the ground commander to be briefed on the situation. The Army had been inserting troops into the area by helicopter when they came under intense small arms fire. They had lost three helicopters, one pilot killed, and had six helicopters damaged trying to extract the troops already landed. The first four air strikes were to assist the troops-in-contact. We were fortunate that I was able to hit the core of the VC with the first set of fighters as this broke their resistance and we caught groups trying to swim back across the river. The fighters strafed many of them on the banks before they could escape among the reeds.

During these strikes I received a call from the Army commander, telling me he had spotted from his command chopper activity 16 kilometres further to the northwest. I already had more fighters available, which I did not require for the troops-in-contact so we headed for the new target. Two of the fighters were flown by Captain Richard C. McNulty and Captain Craig R. Iverson from the 120th Tactical Fighter Squadron. Iverson later put me up for a gallantry award but watching their attacks, I believe they were the gallants. My next set of fighters were two A37s. As the number two aircraft pulled off the target he called that he had been hit in the leg but was all right. I saw him on his downwind journey so I marked a target for him again, and cleared him 'hot'. When I looked back I could not see him and the lead aircraft could not raise him on the radio. I heard a couple of days later that he did not get back and was listed as MIA. This incident cut short the mission so I had a few minutes to call Cai Be artillery to put in some artillery while waiting for further fighters. With two air strikes and the artillery I was able to confirm 12 enemy killed.

As I was now critically short of fuel and out of smoke rockets I returned to Dong Tam only to be told to return to the same area and put in two more air strikes. Being quite familiar with the area by this stage I already had some targets in mind before I arrived there and we destroyed seven bunkers.

Due to recent night losses every effort was made to have two FACs on each night flight so I teamed up with Don Washburn. We would take it in turns—one flying and doing the usual FAC work while the other monitored from the back seat and looked after the radios. The first night, 12 July, we were following this new procedure I took off into the north calling the artillery to hold their fire for two minutes. As I flew over the silent guns I had a magneto failure and the engine started running like a hairy goat. The black void in front was totally uninviting to consider a force

ENGINE FAILURE

landing straight ahead, which was the normal procedure. As I had partial power, but with failing airspeed, I did a wing over turn back through 180 degree to land downwind on the runway. Just as I arrived back over the artillery, the two minutes was up and they started firing again. We were 50 feet (15 metres) above the 155 Howitzers and 8-inch guns and our cockpit was being lit up by the muzzle flashes. The noise was deafening, the terror immobilising, and I had trouble concentrating on getting the aircraft down. Don was in the back seat screaming over the artillery frequency to stop firing. As I touched down I was conscious of the noise from the guns stopping. I shudder to think how close those artillery shells must have been to us. We landed fast and the PSP was slippery so I had to do some heavy braking, which was not helped by my knees shaking. Neither of us could stand up initially when we shut down the engine. I had trouble talking while trying to explain to Jake what had happened.

Jake summed up his assessment with his usual response.

Safely back on the ground, it took only a few minutes to rectify the problem, which was an electrical lead that had come loose from one magneto. Minutes later, we were back off into the darkness to put in two air strikes, controlling two sets of three F100s and one set of three VNAF A-1s.

Early in the morning on 14 July I assisted troops-in-contact at seven different locations around Snoopy's Nose. Besides six air strikes I also directed artillery from Cai Be. One of the ground units was the 3/39. It was normal procedure when fighters were cleared off target, particularly with troops-in-contact, for them not to fly over the target area. This was done to prevent hung up ordnance falling on friendly troops. When CBUs are laid down, sometimes not all the canisters leave the tubes, but dribble out after the attacking pass.

After I had cleared a pair of F4s off target, unknown to me they did a climbing turn over the contact area. They were probably interested in the amount of smoke and hoped to get a glimpse of the following fighters in their attacks. As I was busily setting up the next fighters, I had a call from the ground troops saying they were being mortared. I asked them from where and was surprised to be told that the mortars were coming from an area where I thought there were no VC. The ground fire I was receiving was coming from an area about a kilometre long situated along a riverbank. The coordinates given to me by the ground troops was a kilometre to the west of that. I sent the new fighters high to hold and flew to the area indicated by the Army but could see no evidence of the VC. I flew low and slow over the jungle but saw nothing unusual and did not receive any ground fire. Rather than waste ordnance I decided to monitor the area and continue pounding the known VC areas. As it transpired, the explosions that fell through the friendly troops were from hung up CBUs from one of the off target F4s. Because of the line of flight of the F4, the Army thought it indicated a direction the suspected mortars were coming from. The ground troops reported taking 18 casualties from the incident.

One of the last F4s I used in the attack said they took a hit through the windscreen but he continued on with his attacks regardless. This indicated there was still resistance from the VC. My last aircraft to attack was an RAAF Canberra, which I had been holding back while I used up the fighters. Although the Canberra was well acquainted with the target by this time, it still took a little extra time for him to line up for his bombing run. During this time I noted some movement on the ground among the devastation left by the fighters. I had the Canberra drop a string of 500- and 1000-pound bombs that nicely churned up the area. After I had completed the fire mission with the artillery, the ground troops were happy to move on and evacuate their wounded.

The following day I visited one of the wounded soldiers in hospital at Dong Tam. He was Sergeant Harold E. Bowman from Weleetka, Oklahoma. His multiple wounds were from the CBU fragments. Sergeant Bowman was sitting up happily in bed. When I apologised for his wounding, he would not hear of it, fully realising it was not my fault. 'Besides,' he gleefully pointed out, 'this is my passport home!' Unfortunately he died the following

ENGINE FAILURE

day when a piece of shrapnel had worked its way into his heart. An inquest was held but no-one was held accountable.

That evening Don Washburn and I went back to Snoopy's Nose and put in four air strikes with eight A37s onto pre-planned targets.

I took Major Nelson out again on pre-planned air strikes on 16 July, which was pretty much a routine affair. The choreography got a little out of sequence at one stage when the fighter pulled off target and turned the wrong way straight into us. There I was, looking straight into the F100's intake as he racked on 90 degrees of bank, reversing his direction. I already had over 90 degrees of bank on myself and was pushing negative 'g' to get underneath him. I got a very impressive close-up view of the underside of the F100, which was pulling vapour trails off the wing tips as the pilot laid into the 'g'.

'Thought I had you for a minute there Tamale', radioed the F100 pilot. As casual as I could sound I said, 'Missed by a mile!'

During the rest of July I flew three or four pre-planned air strikes each day or night. I was clocking up the flying hours quickly doing over 100 hours each month. Thirty percent of my flying was during the night. On one flight I dropped into Bien Hoa for maintenance. I always looked forward to going there as is meant a bit of a break from my normal routine and some pleasant company with the fighter pilots. I did not know that Captain Dick Botesch, whom I had met in Thailand in 1965, was flying F100s in Vietnam. I learned that he had been recently killed when his aircraft dived straight into the target on a strafing run. He was carrying a *Time/Life* photographer in the back seat. It is thought that the photographer got his camera jammed behind the control column and Dick could not recover from the dive. This war had claimed another great human being.

By the time I got back to Dong Tam it was nearing midnight. I was just preparing to start my descent into Dong Tam when, over the village of Ap An Hoa, the deafening noise of artillery started erupting around me. I could see the flashes of the artillery guns coming from My Tho. The ARVN were at it again! As I could see the flashes ahead and the shells were exploding behind me, I knew they were flying through my air space. I slammed the throttle fully open and, disregarding the maximum limiting speed of the O-1, descended steeply at right angles to the artillery fire. At 200 feet (60 metres) above the dark jungle I levelled off and headed for Dong Tam. In just five minutes I was on the ground. The midnight air was cool after recent rain and the only sound I could hear was the sounds of artillery from My Tho like distant thunder. What extremes we experienced in this country!

While doing a VR mission on 22 July along the Thap Muoi Canal, 21 kilometres northwest of Dong Tam, I noticed several new graves in an old cemetery. I descended to low level for a closer look as the situation looked suspicious to me. There had not been any enemy action in that area for a few days and there were no longer any civilians in the area, so it was a free-fire zone. I fired four Willy Petes into the graves and got a small secondary explosion. This confirmed my suspicions that there was ammunition buried there and not bodies. I called for an air strike but there were no fighters available to fit into my endurance time. I tried putting in some artillery but hitting such a small target was too much hit or miss—mainly miss!

On 29 July I found some unusual marks on the bank of a stream in the free-fire zone and on doing a low pass I could see one end of a few sunken sampans sticking out from under the waterlilies. I called for a fire mission from Dong Tam artillery but the water was too churned up when we were finished to see if we had destroyed the sampans.

My civilian pilot's licence medical was overdue for renewal so Major Nelson kindly allowed me to take '981 across to Vung Tau, where I could do my medical with the RAAF doctor, Doctor Tibbett. I would stay overnight there and fly on to Bien Hoa for maintenance the next day. At Vung Tau I had a good meal in the officer's mess and a comfortable bed for the night. While flying on to Bien Hoa I was diverted to assist a downed VNAF O-1 in the Rahng Sak, 100 square kilometres of swamplands situated along the Nha Be River, midway between Saigon and Vung Tau. It had

ENGINE FAILURE

once been a tangle of mangrove swamps and a haven for the VC but the constant dropping of Agent Orange on the area had defoliated most of it, so it was now a barren wasteland of mud and waterways. The VC were still there, living in tunnels burrowed into any portion of land possible, even if it was only millimetres above the high-water mark. The Mobile Riverine Force did magnificent work there, flushing out the enemy under such arduous conditions. When a marine craft landed, the troops would disembark and find themselves sinking up to their waists in stinking, clinging mud. I cannot even imagine what it must have been like fighting under those circumstances. When the 'Ranch Hands' (C123 spray aircraft) dropped their lethal loads of Agent Orange it could be smelt miles away. In 1993 I flew over the Rahng Sak with Air Vietnam and much of the area had since been re-developed into fish farms. After seeing that, and with my knowledge of what took place 25 years before, I did not eat fish again while in Saigon.

The downed O-1 was at the coordinates YS 026 525 but I couldn't see any sign of the crew or any marks in the mud suggesting they had left the aircraft when I arrived. The aircraft had impacted at a shallow angle with the left wing bent downwards and the engine buried in the mud. There were no skid marks so the aircraft must have had very little forward motion when it crashed. I assumed the pilots were still in the aircraft. The two F100s he had been working with were still circling overhead but getting short on fuel. The VC's position had been compromised so they had nothing to lose and were firing at the flying aircraft and the crashed O-1. Not far from the O-1 was a bank of raised ground that did not look as wet as the surrounding areas and it seemed most of the ground fire was coming from there. Due to the fighter's fuel shortage I decided to have them expend their ordnance in one pass. The lead aircraft placed his napalm nicely but it did not spread due to the mud. His number two rolled in, placed his ordnance accurately but as he started to pull out of his dive he transmitted something incoherent. He obviously had his transmit button pressed and I could hear his heavy breathing as he sucked oxygen through his mask. The aircraft had condensation streamers coming off the wing tips as the pilot pulled hard on the control column. It contacted the ground in a wings level attitude and bounced without breaking up. At this stage I expected to see the pilot eject but watched in horror as the fighter continued for perhaps 500 metres before cartwheeling. There was no traditional fireball and the aircraft continued to disintegrate as it cut a trail of destruction. Finally the fuselage alone tumbled sideways across a small stream and came to a halt. As I circled overhead I could clearly distinguish the shape of the fighter's fuselage and the cockpit area. By this time the lead F100 had to leave due to fuel shortage. All I could do was report what I had seen to Tamale Control and head for Bien Hoa, feeling very depressed. After delivering '981 to maintenance I did a quick look around the base exchange and then headed for the fighter pilot's hooch. As usual I was fed a nice steak and provided with a bed in an air-conditioned room. It was a nice finish to an otherwise rotten day.

In the morning I hung around maintenance waiting for '981 but they found a few things needed further attention on the aircraft and I flew a replacement aircraft back to Dong Tam.

The afternoon and night of 2 August turned out to be very hectic. I did three pre-planned air strikes with Major Nelson at WS 984 510, 15 kilometres west of Kheim Ich. The mission was fairly routine but good practice for Major Nelson. We destroyed six sampans and 13 bunkers at the western end of the VC Panhandle and were back at Dong Tam for the evening meal. But at around 11 p.m. I was woken by Sergeant Cover. Troops were in contact back in the vicinity of where I had been earlier in that day. As I was starting the engine, Dick Nelson appeared out of the gloom saying, 'Don't leave me behind!'

Approaching the target area I could see the flares were already dropping and Andy Anderson was controlling an AC47 gunship firing Gatling guns into the night. The spray from the snaking red tracers reminded me of the water from a garden hose being spread over plants. The downward red tracers were mixed with green tracers flying skywards and the parachute flares drifting earthwards.

ENGINE FAILURE

It was like a fairyland of lights in the black night. Andy Anderson quickly briefed me on the situation and departed for Dong Tam, leaving me to continue the pressure on the enemy positions.

Due to the wind shear I was having trouble keeping the target area satisfactorily illuminated for the fighters and they did a number of aborted attacks when the flares blinded them. This turned out to be very time consuming and I lost track of time which led to another problem. The aircraft I was flying had a fuel gauge problem—the tank would be empty but the left gauge indicated in the reserve. The right fuel gauge was fairly accurate. I had been changing tanks from time to time to keep the usage as even as possible but at a very inconvenient time the engine cut dead. I chopped the throttle, switched to the right tank and selected the fuel boost pump. The engine thankfully burst into life again. When I looked at the left fuel gauge it indicated the tank was about an eighth full but the needle was rock steady, suggesting no contents. The right fuel gauge was also indicating an eighth but the needle was bouncing off zero. All this happened as I was clearing the last fighters off target. I had to bid the ground troops a very hurried good night and headed for the nearest airfield, which was the Army base at Can Tho.

During that hasty flight I think I looked at the right fuel gauge every ten seconds as it bounced off zero. As I approached Can Tho I informed the tower that I could not see their runway lights, just a sea of black and I was getting apprehensive. The tower informed me they were under mortar attack and the runway lights were off but there was one flare pot at the centre of the airstrip, off to the left side. I was desperate to get down and I was committed to Can Tho. I had only landed at Can Tho once before, and that was during a rainstorm at night, so had no idea of the ground layout. After visually picking up the solitary flare pot from two kilometres out, I set up my approach, guessing where the end of the runway was. Approaching what I thought was the base perimeter I switched on my landing lights only to see a bunch of coconut palms and no airstrip, and I also attracted ground fire. Off with the lights! I gave it a couple of seconds more and flew at a point just short of the flare pot. Although I was forfeiting valuable runway I figured it was better to run off the far end at low speed than to hit the near end obstacles at flying speed. It was not a tidy landing and the braking was heavy but I was down in one piece. I turned off the runway, not knowing if I was going to roll into a ditch, shut down the engine and we both ran for the cover of a dark structure. The incoming rounds stopped but we sat in our safe haven until we could see ground personnel moving about. We started to survey our surroundings. To our shock we noticed that what we had been hiding behind were 44-gallon drums of aviation gas! We both looked at one another, saying nothing. Sheepishly we returned to the O-1, refuelled and flew back to Dong Tam, arriving by 6 a.m.

The next few days were devoted to ALO duties and routine bunker busting, putting in an average of four pre-planned strikes each day. On one occasion I was sent out to fly cover for a Riverine hydrofoil that had got stuck on a bank of the Xang Canal. These hydrofoils were heavily armed and were ideal for negotiating marshland in the wet season. Occasionally the pilot would underestimate the size of an obstacle he was passing over and get stuck. At these times they were sitting ducks. I had watched them travelling at high speed down narrow canals and virtually pass over Vietnamese in their canoes. I hate to think of what went on in the mind of these poor villagers as these monsters bore down on them.

On about 12 August the 5th/60th were patrolling Highway 4 and I flew cover for them. I spent most of the time at zero feet ahead of them, just in case of an ambush. It was rather irresponsible but enjoyable running along with my wheels just above the bitumen, watching the locals fall off their bicycles and scatter. The only things thrown up at me may have been a few local vegetables! The following morning at 1 a.m. the FAC hooch was targeted with mortars. I am fairly sure that it was not retribution for my low flying! The first mortars hit the base exchange and were walked up to the FAC hooch. Bearing in mind that we used to sleep through nightly artillery barrages, I did not hear the mortars.

The others had scrambled into the bunker but as I had not also appeared in the bunker, they feared the worse. When they checked on me later I was still in bed. My mosquito net had collapsed about me but what did ultimately wake me was one of the other FACs pulling at the mosquito net to untangle me. The sandbags around our hooch had taken the impact but anything above that level was severely damaged. Only that afternoon I had bought a brand new B4 kit bag from the base exchange and it was sitting on top of my locker. It was heavily perforated. I later took it back to the base exchange and claimed I had bought it like that and had not noticed. They kindly exchanged it for another one, as there had been a lot of stock written off in the attack. Worst of all, our air-conditioned room had been badly damaged. A fruit cake someone had acquired was spread up the wall and across the ceiling. It was a few months old but we had not minded scraping green furry mould off it before we savoured a piece, until now.

Mortar attacks on Dong Tam were frequent but the VC would only lob in about six rounds and disappear, before we would get a location on them and respond with our artillery. To the west of the base was a large area, ten klicks by eight klicks, which was a no-fire zone. We told the Army that we had been seeing signs of enemy activity in the area and requested air strikes but they were never approved. As the Army did not seem interested in what we had to say at morning briefings, Major Walker tried to approach General Ewell direct. He was told to go through the chain of command, which he had done but got nowhere. Bill Walker asked me to put on my Australian flight suit and see if I could get in to talk with the general. Although I had heard that General Ewell was a tyrant who did not suffer fools or incompetents and 'ate majors for breakfast', I agreed to Major Walker's request. Besides, I wasn't a major!

As I walked to the 9th Division Headquarters I worked out what I was going to say. I was going to be confident, firm and positive. I was looking forward to meeting this 'legend' as I had read of some of his daring exploits parachuting into Normandy, Arnhem and the siege of Bastogne during the Second World War.

On entering HQ I got past a clerk who, leaping to his feet and saluting, had no idea of what, or who, I was. I was in General Ewell's outer office before I knew it, where a colonel indicated that I could not just march in and talk with the general. Looking past him I could see the great man himself in his varnished wood-panelled office with crossed flags. He saw me, smiled and said, 'Hi ya, Cooper. Come on in. What can I do for you?' Now that I was standing before the general, not knowing how he knew my name, my confidence was waning. My carefully rehearsed speech degenerated into a muttering, stammering mess. I told the general of the concerns of the FACs, which he listened to politely and then he said, 'Thank you, Cooper. I'll look into it'.

Back at the hooch we were all convinced that nothing would happen. True to his word, General Ewell got back to us with an answer. TACAIR and artillery had been requested on the area many times but approval never came. As there was plenty of other work to be handled, no-one ever thought to query why approvals were not coming through. The normal procedure was to send the request to Saigon. Saigon in turn would obtain a clearance from the province chief. Only then could the strikes be launched. The province chief was supposed to be warning civilians of the pending strike but, of course, this was tantamount to warning the enemy. In fact, it was widely thought that the province chiefs were 'double dipping'. That is, receiving payment from the United States for bombing his territory and being paid again by the enemy for letting them know what was coming. In this case, the province chief had been dead for six months. Little wonder the approvals were not coming through. The problem was soon rectified and, lo and behold, the nightly mortar attacks on the base reduced as the VC hideouts were pounded with artillery and air strikes.

Major General Ewell tended to be a little unpopular. Author and leading military analyst Colonel David Hackworth always wrote about General Ewell in an uncomplimentary manner. This is not surprising as General Ewell was at the top and his commands could lead to soldiers being killed. That is war. General

Ewell expected only the best from his men and expected complete discipline. Colonel Hackworth did not agree with many of General Ewell's methods but, instead of diplomatically working around his commander's rules, he went at full confrontation. Both these men were highly experienced combat soldiers and way above my league so I cannot enter into their argument. I did hear that Colonel Archer, CO of the 2nd Brigade, had also crossed swords with General Ewell on the subject of 'body count' being used as a gauge on the success of operations against the enemy. Soldiers are given guns to shoot the enemy, which leads to 'body count'. Take the gun away from the soldier he will then use a club, or even his hands, to dispose of the enemy. If the soldier kills every enemy, that must lead to 100 per cent success. It must then follow in percentages down from there. Get rid of the enemy and the war is won. This is simplistic but, in a war such as we experienced in Vietnam, I could see no better way of gauging success. Personally I found General Ewell to be an officer of great compassion and concern for his men. The fact that he has pursued the Medal of Honor recommendation he submitted on me for over 36 years is an indication of his compassion and amazing tenacity. He is a total gentleman and I would fight a war under him again any time.

On 13 August I was given some pre-planned strike coordinates that were along the northern bank of the Vam Co Tay River. The only object of significance anywhere near these coordinates was an abandoned, large old French colonial mansion situated on a disused copra plantation. The immense roof was constructed with terracotta tiles, an expensive product nowadays in the Western world, and the overgrown gardens led past marble statues to a large entertaining wharf area, all cut out of solid stone. It was a relic of French decadence. The fighters I was allocated were two F4s that had been diverted from a target in North Vietnam due to bad weather. They were carrying 2000-pound slicks. This was a size I did not see very often in the south. As the mansion was derelict and isolated I did not have to mark the target but merely indicated to the fighter what they had to bomb. It was awesome to watch those bombs sail all the way down from the release height of about

4000 feet (1200 metres) and go through the roof, leaving two neat holes. I thought for a moment that the bombs were duds due to delayed fusing. Then, as if in slow motion, dust and debris appeared out the doors and windows, the walls disappeared and the roof fell into the mess. By the time both F4s had finished, the once proud two-storey mansion and surrounding buildings were nothing but dust. If there had been any VC in the basement of the building, they would have been dust as well.

About midday I was looking over the intended target at XS 190 451, just west of Cai Be, when a pair of F4s checked in on my frequency. I briefed them about the series of concealed bunkers we were aiming for. As we were about to lay down the ordnance another pair of F4s checked in. The lead aircraft was being flown by Captain Anthony Sultan of the 559th Tactical Fighter Squadron. His call-sign was 'Phantom 21'. He told me he had a fuel problem and asked if I could expedite his mission. As I had seen a few concealed sampans nearby, I decided to run both sets of fighters together as the first set of fighters were also short on fuel. The cloud base was at 2500 feet (762 metres), which would make it quite restrictive for delivering slick bombs. I set up the fighters in reciprocal patterns, one left and the other right. The targets were about 500 metres apart and at times there would be two fighters, from different squadrons, on their attack runs almost in formation, with me in between. I alternated them between targets and was pleased how well it worked out. The fighter pilots were very impressed with my performance and I received a Letter of Commendation from the 12th Tactical Fighter Wing. This is a technique that I would never have attempted with troops-in-contact. The F4 pilots' professionalism and ability were a large part of the success of this mission.

A day or so later I was waiting for my aircraft to be refuelled, watching the helicopter movements across the airstrip. About 100 metres away was a Chinook that was being refuelled while its engines were still running. I was thinking about how unsafe this was when I noticed the helicopter start doing an unusual bounce on its oleos. Within microseconds the whole aircraft started to

ENGINE FAILURE

bounce wildly and spun through 90 degrees. The two rotors smashed together, tearing out the front engine. One rotor cut off the arm of the crew chief refuelling and the front rotor chopped the cockpit in half, killing the two pilots. The smashed Chinook fell into a heap with bits flying over a large radius. I stepped back inside the revetment as a hydraulic jack bounced past me. I had heard of ground resonance associated with helicopters but did not realise it could be so devastating.

With the chopper lines close to and along the length of our landing strip, the choppers used to cause a lot of turbulence with their downwash, particularly with a crosswind. Sometimes the turbulence made the little O-1 almost uncontrollable. When we were landing and we could see choppers with their engines running, the FACs would ask them to go to ground idle. In ground idle the chopper was particularly susceptible to resonance and the chopper pilots would be sometimes very testy about our request. After witnessing the Chinook disintegrate I seldom asked them to go to ground idle any more.

After doing a VR mission I recovered to Bien Hoa for maintenance checks. When I arrived back at Dong Tam at about midnight the base was under attack again so I diverted to nearby Truc Giang. The airstrip there was constructed of rubberised PSP matting. When I touched down it was so smooth it felt like a cat pissing on silk, so to speak. I taxied to the refuelling area and there was not a soul in sight. The place was in complete darkness and I had an uneasy feeling. I could hear Vietnamese voices in the distance, so after quickly refuelling I got airborne again, climbing to 10 000 feet (3000 metres) deciding to spend my time waiting in the air. It was beautiful at that altitude. The air was cool and smooth and I could just make out the lights of Saigon in the distance. Everywhere else was basically the black of the night. As it transpired, within 30 minutes I had clearance to return to Dong Tam but I found out there was another series of enemy mortars inbound, impacting into the far eastern side of the base. I could see the FSB returning artillery fire. Rather than disrupt their fire mission I elected to land downwind where I did not have to ask them to hold their fire as I passed overhead on approach. As I crossed the My Tho River I emptied my eight smoke rockets into the bank short of the base to discourage any would-be assailant.

The Commander of the 3rd Brigade, Colonel Benson, completed his tour of duty on 15 August and was replaced by Lieutenant Colonel John Hemphill, who had been with Warrant Officer George Gray when he was killed in July. It was not too long before Colonel Hemphill was shot down and wounded again.

The following day I was putting down an air strike at XS 320 452 when I noticed a group of manned sampans tucked in under some vegetation overhanging a small stream. The F4s I was directing had just completed their last run so I called Dong Tam FSB for a fire mission. As I could see no peasants in the area, I had the FSB fire for effect with VT Fusing. After the fire mission I could see six VC floating in the water. Before I could do anything further I received a call notifying me of troops-in-contact at XS 888 520. Two very accurate air strikes had the Army up and moving again. At about 3 a.m. I was scrambled back to the same target as the previous afternoon, where the same Army unit was in contact with the enemy again. I counted only six tracers fired in my direction and the Army were content with the results of one air strike. I began to really like the easy ones.

On 17 August I put in one air strike using the B57. The B57 was an American derivative of the Canberra but they could dive-bomb. Due to their size, the B57s appeared to be going very slow but I was impressed with their accuracy. It was a pity to waste their accuracy and bomb load on busting bunkers and clearing trees.

9

DOWN IN THE BOONIES

AT ABOUT 4 A.M. ON 18 August Mike Cover, the radio operator, woke me up. I had been flying the night before until midnight so it took some shaking to bring me round. The sergeant was saying, 'Sir, Sir, all brigades [1st, 2nd and 3rd] are in contact with the enemy and you are rostered as air liaison officer.' As he was trying to wake me, I could not fathom why he spoke in a hushed voice. I grunted something incoherent in response. Then the sergeant added, 'You are to meet the 1st Brigade commander at the chopper pad and take your grunt gear as you may be gone for a couple of days.'

The hooch was in darkness so I had no idea if any of the other FACs were in their bunks or out flying. It took me about 5 minutes to get my meagre gear together—I instinctively knew where everything was so it was unnecessary to turn on any lights—and stumble outside to my transport. The morning air had quite a chill to it and I felt cold as the sergeant drove me in the open jeep to the helicopter pad. A life of continuous combat tended to cause you to become something of a robot so I just sat there not even thinking of what was to come next. It was not necessary for me to quiz the sergeant about further details as I had learned that this did not bring a flood of information from him. Each person in the chain was told only enough to get him moving in the right direction and then he was expected to handle what followed with his initiative, ability and what he had learned from bitter experience.

The sergeant dropped me at the helicopter pad saying, 'Good luck, Captain'. The Americans generally referred to me by the equivalent US rank or whatever they deemed appropriate at the time. I checked several helicopters that looked like they could be my ride but each time I was met with a shrug. After about 15 minutes of this I saw a Raven Hiller OH23 arriving, but not shutting down, some distance away near the MRF base so I ran down to check it out.

The Raven was manned by a full colonel in the left seat and an Army warrant officer in the central pilot seat. In this aircraft two passengers sit, one on each side of the pilot. I did not recognise the helicopter or the pilot. Most other times I flew in 3rd Brigade OH23s. On the front bubble of these their radio call-sign and unit crest was painted. The call-sign was 'Merkin'—Webster's dictionary describes it as a false pubic hairpiece! The unit crest depicted what looked like a psychiatrist's smudge with a pair of wings attached. The whole picture purposefully resembled an abstract view a gynaecologist would have of his patient.

I approached the passenger in the left seat asking, 'Are you Colonel Archer, Sir? I am Flight Lieutenant Cooper, your ALO.'

'Jump in, Cooper,' he replied. As I strapped into the right seat I said 'Hi' to the pilot and he acknowledged with a handshake and a smile. I was still strapping in when the colonel said, 'Let's go, let's go', upon which the pilot manoeuvred the machine into the air and headed off at about 50 feet (15 metres) above the ground, while I worked out how to plug the electrics into my helmet and strap in. Our destination was Can Giuoc to the east of Tan An, about 20 klicks south of Saigon city. This was the base of the 3rd/36th Infantry, US Army. On arrival, the pilot flew at zero feet and about 50 metres in front of the firing artillery to stay below the line of fire. I always found this particularly unnerving as the artillery fire was deafening and, being only a passenger on the aircraft, I felt quite helpless and not in control of my own destiny. On landing Colonel Archer said, 'Don't go away, Lieutenant.'

As the other person present was a warrant officer, I assumed he meant me. In any case, where was I going to go? This was the beginning of many different references to my rank throughout the day. Colonel Archer referred to me as Cooper, Lieutenant, Captain, Major and Son, probably because he did not understand the Australian rank insignia I wore on my shoulder. This would cause me added confusion as I could never be sure he was talking to me. As soon as the Colonel disappeared around the corner, the warrant officer, Mister Jones—warrant officers in the US forces are all called Mister—said, 'See ya shortly', and also disappeared, leaving me standing at the helicopter to amuse myself. This went on many times throughout day, making me feel rather unneeded and a bit like the 'Invisible Man'.

After about 20 minutes both the colonel and Warrant Officer Jones appeared in something of a hurry and, without a word, we re-boarded the Raven and took off below the firing artillery. We headed back to the west towards Binh Phuoc where I gathered, by listening to the radio transmissions, that an element of the 3rd Brigade were in contact with the enemy just north of the base. Binh Phouc I remembered well from the previous April when I was called to help the 2nd/4th Artillery, who were under attack by a large force of Viet Cong. After a lot of Army talk the colonel asked me on the intercom, 'Captain, do you think we can use TACAIR on this target?' I answered, 'Yes, Sir. I can have it for you in about 20 minutes.'

'Arrange it,' he replied. He then went back to 'Army speak' on his radio.

I called Tamale Control to arrange the TACAIR on the VHF set. 'Tamale Control, this is Tamale 35 on "Victor".'

'Go ahead, Tamale 35,' came the instant reply from Sergeant Cover. I admired the efficiency of the radio operators. You never had to wait for a reply, they were always right there when you called, day or night. I pressed the transmit button saying, 'Tamale

Control, we need a FAC and TACAIR immediately just to the north of Binh Phouc.'

'Standby, Tamale,' was the response. It was now only two minutes since Colonel Archer instructed me to arrange the TACAIR and he was back on the intercom asking, 'How are you doing with the TACAIR Lieutenant?'

'On its way, Sir', I said, trying to sound confident but not really being sure. Just then Tamale Control came back saying, 'Hello, 35. Tamale 32 is in your area and will be with you in five minutes and I can get one pair of F100s there in 15 minutes followed by two more if required.'

'Roger, Control. Have Tamale 32 call me this freq.'

'Will do.' One minute later I hear Tamale 32, Captain Don Washburn, on the air.

'Tamale 35, this is 32.' I reply, 'Hello, Don. We have troops-in-contact five klicks north of Binh Phouc, contact "Big Trooper" on FM 89 for briefing.'

'Roger that, 35,' comes the brief reply. The Americans are not ones for the regimented radio procedures the Australians use. However, it works and I never once became confused as to who was talking to whom on the radio.

'How's the TACAIR going, Cooper?' asked the colonel.

'All organised, Sir.'

'Good work, Lieutenant.' I will have to remain alert here I tell myself.

While I was busy on my radio, the colonel had changed frequencies and was now talking with Tan An, home of the 1st Brigade. The next moment we were back under the Tan An artillery firing at the contact north of Binh Phouc and I was terror struck but maintained an outward appearance of nonchalance. On the ground Colonel Archer took off in one direction and Mister Jones another. This time Mister Jones said before departing, 'Would you supervise the refuelling, Sir?'

'Sure thing,' I replied but thinking it was not really my job.

I sat there with my fingers in my ears for 20 minutes to dampen the constant deafening thunder of the artillery. Mister

DOWN IN THE BOONIES

Jones beat Colonel Archer back to the chopper pad. I gained the impression that he had only recently been transferred to another unit and was taking the opportunity to see various acquaintances from his old unit on these landings. Colonel Archer duly arrived back announcing, 'We're off to HQ Saigon.'

The flight up to Saigon was much more enjoyable, cruising at 1500 feet (450 metres) above the small arms envelope. Without side doors on the Raven I was actually starting to feel cold with the refrigeration effect of the slipstream on my sweaty flight suit. At Saigon we landed on an old racecourse or sports field in the south of the city and, once again, the Colonel and Mister Jones headed off in different directions. 'Back soon, Captain,' said Colonel Archer. Would I ever get used to these promotions and demotions!

At Saigon there were no guns firing and it was rather peaceful sitting by the chopper with my flight suit wound down, catching some sun. It was mid-morning and I had not eaten since mid-afternoon the day before. I was getting rather hungry and thirsty. Nearby I saw some Doughnut Dollies serving coffee to APC crews. I wandered over and became a point of interest with my strange uniform and accent. The girls kept giving me biscuits and coffee while they asked me all about Australia, vowing they were going there on R & R. This time I had decided to take care of the refuelling so I reluctantly severed myself from the female company. They were always very friendly, helping you to realise what life was really all about. Not seeing white women for several weeks at a time had one emotionally disappearing into a strange, unreal word.

When Colonel Archer arrived back he was in a hurry and my heart sank as I detected that we were off to another troops-in-contact—the 2nd Brigade were fighting north of Cai Be, west of Dong Tam and 60 klicks from Saigon. The flight to Cai Be was at high speed, if you can imagine a Raven going at high speed. Colonel Archer's Assault Support Patrol Boats and Monitors were engaging the enemy north of Snoopy's Nose, a well-known hot spot for the Viet Cong. I could never understand why we kept going back there with engagement after engagement when the whole area could have been defoliated to deny a hiding place, or just simply annihilated with B52 strikes. There were not only boats of various types involved, but also foot soldiers in a number of different locations, all in close contact with the enemy.

Colonel Archer said to me, 'How about TACAIR on this one, Major?' Another promotion! I assessed the contact as too close and too complicated for TACAIR, so I said, 'Colonel, it's too complicated but I can do a better job than the grunts are doing with the artillery.'

'Go ahead, Cooper. The aircraft is yours.'

I had to then sort out with the ground commander where he wanted the ordnance and where the friendly troops were. Whoever had been directing the artillery had been lobbing it short with a good chance of hitting friendly forces. As the artillery was already on station, it simplified my procedures and all I had to do was readjust the ordnance. The Colonel insisted we stay below 500 feet (150 metres) as that was the perspective with which he was familiar. I would have liked to have gotten a bit higher so I could have a better overall view but in this case I was familiar with the territory—I had put in air strikes in this immediate area before and would do the same on many occasions to follow.

As the Viet Cong knew their positions were compromised, they had no tactical advantage to lose and hence they fired upon everything that moved. Flying low made us a prime target and we started picking up heavy automatic weapons fire, taking a couple of hits. However, now that I was preoccupied with my task, I took little notice of the ground fire. Earlier, before I was given control, I had the usual reaction crouching down as small as possible. After about 20 minutes of artillery adjustment the ground commander was happy to lift the firing and I noticed the ground fire on us came to a stop.

By this time we were getting short of fuel and Mister Jones suggested we head for Cai Be. I was not happy to leave the target area as I felt our presence might still be required. This I conveyed to Colonel Archer by saying, 'I feel we should get back here as soon as we can Colonel.'

'OK, Lieutenant', came the reply.

We landed by the river in Cai Be where there were some MRF boats. Colonel Archer said, 'I'll be right back', and disappeared toward the boats. This was the first time Mister Jones did not drift off as well. While he was refuelling I asked, 'How long have you been flying, Mister Jones?'—he looked rather young.

'Twelve months, Sir, and I have been in country for six.' I was entrusting my life to a novice!

'And how old are you?' I enquired.

'Twenty, Sir.'

I clearly remembered how overconfident I had been at that age and shuddered inwardly. Mister Jones told me that he had just been married, or was about to be married (I do not remember which). After this short exchange, he announced, 'Must get some shut-eye.' He slumped in the left seat, crossed his arms, pulled his peak cap down over his eyes and went off to the land of nod. He was not being unfriendly, just practical. I knew if I closed my eyes I would only feel worse with the little sleep I had had in the past 24 hours. So I was on my own again.

It wasn't for long as the colonel was soon back saying, 'Let's go'. We took off and headed back to the Snoopy's Nose contact. When we arrived we spent 45 minutes just circling around and talking with the ground commander on the radio. There had not been a further shot fired the whole time and the area appeared secured, so the colonel bid farewell to the ground commander and said to Mister Jones, 'Let's head east along the Mekong to Rach Kien.'

'Roger that,' said the pilot and, much to my horror, away we went at 50 feet (16 metres) again. I said to the Colonel, 'Don't you feel that we would be better off at a higher altitude, Colonel?'

'See much more down here,' came the reply. I had to disagree but he was the boss. We headed east past Dong Tam and at My Tho headed northeast back over Can Giouc to the Vam Co Tay River.

Mister Jones asked, 'Are we going to Rach Kien or nearby Colonel?'

'Nearby,' said the Colonel. 'I have some boats on the Rach Cac River supporting the 1st Brigade.'

'Then, I am going to need to gas up,' responded Mr Jones, 'Just in case.'

'I have a refuelling barge at the junction of the Rach Cac and the main river,' said Colonel Archer.

'Roger that,' was the reply.

The refuelling barge was the smallest piece of equipment I had ever seen. That it was intended for an aircraft to land on it seemed incredible. It was quite scary for me to stand by and watch a 20-year-old controlling his machine down onto this 'iddy biddy' postage stamp. It was barely large enough to contain the helicopter skids. I couldn't imagine that they also landed the Hueys on this type of devise. However, land he did—perfectly. The pilot shouted out to the barge crew, 'Fill 'er up!', while we remained strapped in our seats with the engine running. The whole operation took virtually three minutes before we were lifting off and away again. I begrudgingly conceded that maybe this kid could fly.

Once airborne Colonel Archer called Rach Kien on FM, which made him decide to go straight to the 1st Brigade contact area rather than the 5th/60th HQ. The contact was about 20 klicks directly south of Saigon. Just south of this was the main landing area where the Hueys were operating. They were inserting troops at the actual contact point with the enemy about two klicks north. Another Raven occupied the landing pad along with a bunch of trucks and APCs, so we landed in deep grass just south of Highway 279, which ran southeast from Rach Kien. I did not see the pilot of the other helicopter. My pilot must have known where the other Raven pilot was because he disappeared as soon as Colonel Archer left the heli-pad.

I wandered across to some nearby Army vehicles. The crews were lazing around and appeared to be just killing time. I wanted to see if I could scrounge a cup of coffee. There were a large number of different types of trucks and other vehicles associated with the 1st Brigade's operations. One of the APC commanders was David Rohe of the 1st Platoon, who invited me up onto his machine for coffee. Along with his crewman, an African–American, we discussed Australia and R & R at great length. Another grunt by the name of Don Goetz joined us. I had forgotten this encounter when I met him again some 30 years later at Byron Bay in northern NSW, where he had taken up residence. After discussing certain places we had been in Vietnam, we both suddenly realised that we had met in August 1968.

My pilot returned to the Raven and I took a picture of the scene not detecting that he was taking a pee in front of the chopper. As it transpired that was probably his last. By this stage it was getting late in the afternoon so we had been on the go for 12 hours and I was feeling quite weary, having had only minimal sleep in the last 24 hours. The discussion with the crew of the APC had taken my mind off my physical state but we were disturbed by the return of Colonel Archer. The Colonel was accompanied by none other than the commander of the 1st Brigade, the legendary Colonel Hank 'Gunslinger' Emerson.

I had not met Colonel Emerson before but had heard a lot about him. He had already been shot down three times. Next month, he was to be shot down again and rescued by Colonel Ira Hunt. Ira Hunt was a 1945 West Point graduate, who subsequently replaced Emerson as the 1st Brigade CO while Colonel Emerson was recovering in hospital from burns. Two crewmembers were killed in that crash.

Colonel Archer pointed in my direction and Colonel Emerson called out, 'Over here, Captain Cooper.' He introduced himself and then proceeded to address me as Garry. It gave me quite a lift to be addressed so personally by such a great warrior. When I had joined the colonels, Emerson produced a 1:50 000 sealed map of the area marked with the location of the friendlies and the enemy. The enemy was the VC 265 Main Force Battalion, which had recently absorbed the 6th Local Force, or the 5th Nhe Be.

'Garry, do you consider we can use TACAIR in this situation?' he asked, pointing to the area on the map. The position of the ground troops was clearly separated from the VC so I said, 'Yes, Sir.

But I would like to view the situation from the air before calling out a FAC.'

Emerson replied, 'I really want TACAIR—you're a FAC, can you do it from the Raven?' I had not considered this procedure but it did not seem to contravene any code I had read.

'No problems there, Sir,' I replied. 'But I would still like to see the area from the air before committing.'

Emerson asked Archer, 'Can we use your bird, Bob?'

Colonel Archer said, 'I do have some more calls to make, how long will this take us, Lieutenant?' I had to sound confident so I replied, 'If I can get some fighters that have been diverted from another target and they have the right ordnance, we can be leaving this area within 50 minutes.' By this time our pilot was within earshot.

'Right!' said Colonel Archer to our group, 'We will get airborne and let Cooper decide whether to put in TACAIR or not. Either way, we will not come back here. It looks like I am not going to get back to my HQ tonight so I would like to spend the night in Saigon.'

After many 'real fines' and other pleasantries we were airborne again. The target was only a minute away so I got on to Tamale Control to see if there were any fighters available while we headed towards the contact area.

'Tamale Control, this is Tamale 35.' Back came the reply on VHF immediately, 'Tamale 35, go.'

'I'm at XS 815 665 with troops-in-contact, do you have any fighters for me?'

'Roger that, Tamale. We have a flight of F4s short on fuel due to a weather diversion from a target up north. Can you use them right away?'

'What are they carrying, control?'

'Wall-to-wall high-drags and 20 mike-mike, 35.'

'Excellent, just what I need. My position is 175/27 off Tan Son Nhut.'

'Roger, 35. Hammer 41 should be with you in five.'

I had just enough time to look over the target. A lot of time was saved here as I had already viewed Colonel Emerson's situation map whereas on an unknown target it could take 15 minutes to get a grasp on where the friendlies and the enemy were. We were now over the contact area, in the vicinity of the village of Ap Tay Phu, and Colonel Archer was in contact with the 2nd/60th ground commander on FM.

No sooner had we arrived on site than we were being fired on. There were a number of other helicopters flying around which our 'boy pilot' was doing a great job of avoiding. I was starting to gain a lot of respect for this young warrant officer. At first I was trying to keep track of all that was going on around me as I would have in my Birddog but there were just too many variables. In the O-1 I would be subconsciously putting the aircraft in the position I wanted. In this case I did not have control of the machine and had to be content with the pilot taking evasive action on other helicopters, trying to avoid ground fire as well as to responding to interruptions by Colonel Archer who would not stop doing his Army thing! Eventually, in fact it was only seconds, I said, 'Colonel, if you want TACAIR I am going to have to have complete control so stop interfering and directing the pilot.'

To my surprise he said, 'Sorry, Captain. Your ship.'

In the middle of this exchange the UHF radio burst into life: 'Hammer check.' Then number two checked in, 'Hammer two.'

'Tamale 35, this is Hammer 41.'

'Hammer 41, this is Tamale 35. Go.'

'Hammer 41 is a flight of two F4s with wall-to-wall Mk82 high-drags and 20 mike-mike, be with you in two.'

'Roger, Hammer. My position is midway between Saigon and The Testicles at 200 feet [60 m] in a Raven helo.'

'We were down here a couple of days ago, Tamale, and know Can Giuoc. Are you near there?'

'Roger, Hammer. Five klicks south.'

'OK, Tamale. We see a lot of smoke and helo activity just to the east of a north/south river.'

'That's us. Standby for briefing.' I then asked the Raven pilot to clear the other helicopters from the area.

'No probs there, Captain,' he replied. One has just been shot down on target and three others have departed with battle damage. If it gets any hotter, we are out of here as well.'

The intensity of ground fire had picked up but I had not been paying it a lot of attention due to being preoccupied with my job. I then asked Colonel Archer to have the ground troops to mark their positions on my command.

'Hammer 41, you ready for briefing?'

'Go ahead, Tamale.'

'Hammer, we have troops-in-contact and your target is an entrenched VC Unit. Friendlies are situated 200 metres east of the target and will pop purple smoke.'

'Friendlies 200 metres east, smoking purple,' he acknowledged.

'Your attack direction will be north/south with right turns and your best bale out will be back at Can Giuoc.'

'North/south, right turns, bale out at Can Giuoc', acknowledged Hammer lead.

'Ground fire is heavy but I can not give you random headings due to the proximity of the friendlies.'

'Understand, Tamale.'

'We will use the high-drags in pairs first, then spread the 20 mike-mike.'

'Roger, Tamale.'

'As we are marking from low level—I will sit west of the target at 200 feet [60 m] after marking.'

'Standby for my mark Hammer . . .'

In mid-sentence all hell broke loose as our Raven was raked by automatic machine-gun fire from our rear left quarter. The pilot had been manoeuvring in a left climbing turn and was about to reverse to the right so that I could put down the smoke canister marker. The aircraft started spinning violently to the left and ascending toward the west. I received a tremendous jolt to the helmet and could see a lot of blood out of my right eye only. As it turned out, a .30 calibre round had hit my helmet, twisting it around so it blocked the vision from my left eye. To me the situation seemed that I had been hit in the left eye and that the blood was mine. This put me in a fatalist frame of mind, which numbed me. Along with the mad gyrations of the doomed helicopter, I was not fully conscious of what was happening. I do recall shouting voices on the radio and glimpses of a river and green countryside, which seemed to go past in slow motion for a long time.

The Raven hit the ground heavily and was still gyrating in the deep grass. All went smooth for a few seconds as we bounced into the air again. By this time my helmet was back in position and I could see with my left eye. The only thing familiar to me in that helicopter was the ignition switches, which I reached across and turned off as we heavily impacted the ground again. The right side of the Raven crumpled and I had the full weight of the pilot on top of me. All went quiet and I just hung there in my safety harness, totally deflated. When I smelt petrol I made a mad scramble to get out through the less than a metre doorway left on my side. The main bubble of the chopper was severely cracked and broken but without large holes. I threw off my helmet, thinking that would make me less bulky in my escape bid, but the weight of the pilot was jamming me against the right side of the wreckage. There was no movement from the other two men, nor was there anything said. For the first time I looked across the cockpit. I could not see the colonel as the pilot was on top of me. The pilot's head was a bloody mess and I was convinced he was dead. I grabbed his dog tags, gathered up all my strength and forced my way out of the small opening onto the ground where I found my legs would not hold me.

As panic overcame me, my first instinct was to run. To where, I had no idea but my legs would not carry me. I propped myself up against the broken Perspex bubble, facing the mess inside. The colonel was slumped on top of the pilot but appeared unhurt. As I stood there I could hear nearby gunfire. I could also hear pinging noises, which I took to be the engine cooling down. Then it occurred to me that both noises were in harmony—the helicopter was still taking bullet hits. Once again panic overwhelmed me and that suddenly got my legs moving. I raced around to the left side of the helicopter where the colonel was seated about waist high. He was starting to move and groan. I undid his safety harness and pulled him out with his left arm around my neck and his M16 jammed between us. He had the M16 lanyard secured over his left shoulder for ready access in flight if he needed it. He fell into a standing position alongside me and I was able to drag him away from the helicopter.

I was totally disorientated but the obvious way to go was away from the gunfire. We were fortunate to be in grass over a metre and a half tall which made us a poor target from ground level. Had the VC been elevated it would have been a different story but they could probably only see the top of the Raven protruding above the grass. I had no way of knowing if the VC were shooting at us or just the helicopter but the low passes by the F4 pilots would have prevented them from advancing towards the wreck. The F4s were not able to fire as they did not know where the enemy and friendly troops were. Further, there was no-one on their frequency to give them direction or clearance. However, unbeknown to the fighter pilots, their low passes kept the VC at bay and prevented us from being overrun.

After stumbling through the deep grass for about 20 metres we came across a low mound that was the western boundary of the disused rice paddy we had crashed onto. I got the colonel behind this low mound, not having the stamina to go any further, and we just lay there. I was wearing a survival beacon vest and attempted calling the fighters but the radio appeared dead. It was the responsibility of the maintenance crews to keep the batteries charged but they tended to be a little slack in this area. Few of the other FACs ever wore the beacon for this reason, so my faith in habitually wearing this uncomfortable apparel all the time proved unjustified and I never wore it again. Later I thought, if the beacon had been working, I would have been able to still give the fighters some direction to deliver fire upon the enemy, which would have eased our predicament considerably. It was unorthodox, but quite possible. After a short time the colonel started to become coherent and said, 'What's the situation, Lieutenant?'

At first he could not comprehend where we were. His immediate concern was for the Raven pilot, who he had noted was not with us. I gave him the dog tags and assured him that it was not in our best interests to return to the wreckage. The colonel accepted this and asked, 'What do you think we should do?'

I said, 'Let's head along this mound away from the sound of the gunfire.'

'Lead the way, Garry,' he said. And to think I thought he did not know my name!

We crawled on all fours along the low side of the mound with the occasional round whipping through the grass near us. As we moved further along the rounds did not follow us. We knew then that the VC did not know we were on the move and, indeed, that there were any survivors. Moving on all fours was a difficult and painful exercise. My AR15 kept getting in the way as we tried to retain a profile within half a metre of the ground. The AR15 swinging around alerted me with a jolt that I had lost my spare ammunition. I always carried a bandolier of six or so, full magazines but must have lost them during the crash or escape. Taking stock of the situation I realised that between us we had only four or five magazines plus six rounds in my revolver and whatever the colonel had in his .45.

After we had gone some distance in this rather undignified manner we heard a Huey approaching and it hovered over the downed Raven. We initially considered returning to the wreck but then decided against that as the Huey was returning fire and any movement on the ground may have received the unwanted attention of the door-gunner. Also, the Huey could have pulled out, leaving us back in the undesirable territory we had just vacated. We tried to get the attention of the Huey, which was at least 150 metres from us, without drawing the attention of the VC toward us. Unfortunately, the tail of the Huey was toward us and I am sure the crew would have been far too busy to be looking at areas away from where the ground fire was coming. There was considerable ground fire being directed at the Huey so we took the opportunity to put more ground between the enemy and us.

At the time it occurred to me that the Huey was spending a lot more time inspecting the wrecked Raven than the situation justified so they may have been extracting the body of the pilot. In 2002, when a leading Washington researcher, Bruce Swander, was investigating the incident, he could not find an Army pilot in that area on that day KIA, even though there were others in different areas and one who died later from wounds inflicted on that day. Bruce Swander came to the conclusion that the hovering Huey recovered the Raven pilot and thus, receiving early medical attention, he survived the war. Thinking back to the condition I last saw the pilot in, I would not have thought this possible. However, at a 5th/60th reunion in Denver, Colorado, in 2001, I had met a Jim Deister. Jim received severe head wounds when Fire Support Base 'Jaeger' was overrun by the VC in February 1968. His head wounds were so severe, he was pronounced dead and had been placed in a body bag. When the medics were unloading the dead back at Dong Tam they found one bag to have an occupant kicking. That was Jim. He was subsequently operated on, medivaced to the United States and, despite being badly scarred, lives happily with his lovely wife Rita in Salina, Kansas. That was a great ending to one hell of a situation.

A further interesting development occurred in 1993 when I re-visited the crash site. At the time it was still difficult to travel around Vietnam unless you were on a police-escorted tour. My first officer, Jeff Bower, and myself got within two kilometres of the site before we were directed by the police to return to Saigon and obtain a permit. A number of ex-South Vietnamese veterans were milling around us like long-lost friends, which may have offended those who had fought with the VC during the Vietnam War. Those who served with the VC were treated well by the government, being given houses and other benefits. Ex-members of the South Vietnamese forces received no benefits and were not treated with respect. The war cemetery for the communists north of Saigon consisted of manicured lawns with white headstones, while the South had an overgrown, ill-kept presentation.

It took us six weeks to obtain the permit and we had to be escorted. Our escorts were ex-VC officers and one was General Thanh Ngoc Nha who had been a colonel on 18 August 1968 and took great interest in my account of the day. He spent more time interviewing me over cups of tea and biscuits than the visit to the crash site took. They told me that there had been ten helicopters shot down in a five-kilometre area and their faces lit up when I mentioned I had been on a scout helicopter, as this revealed to them which crash it had been. When Thanh and his soldiers found the helicopter in August 1968, there was no-one in it and they had no idea that Colonel Archer and myself had escaped to the north. This indicated that the chopper hovering over the wreck had extracted the pilot or his body. Standing at that site in 1993, with a cool breeze in my face and surrounded by serenity, contrasted sharply to when I had last been there. I would really like to know what happened to Mister Jones, a young man of just 20 who had his life still ahead of him.

As the Huey pulled away from the Raven it accelerated to the south, away from our position. We abandoned all hope of getting its attention. Once again I had a feeling of despair as the sound of the Huey faded in the distance, leaving us isolated. I am sure Colonel Archer felt the same as it reflected in the tone of his voice when he said, 'Let's move out'. When we considered we were far enough away from the crash site, and out of the sight of the VC, we started walking in a low, hurried crouch, which still kept us hidden within the tall grass. Occasionally the colonel would stumble as he was still unsteady on his feet and the crouched motion was not easy to maintain. When he fell, I had to assist him back into motion as he seemed to not want to move once he was down. Finally we came upon a small stream overgrown with reeds and full of foul-smelling stagnant water. Colonel Archer slipped into the reeds and said, 'We will hold up here.'

I said, eyeing the foul liquid he was lying in waist-deep, 'Don't you think it would be better to make our way east along this stream to the main river—we might then be able to signal one of your boats.'

'Out of the question,' said Colonel Archer. 'The banks of the main river will be thick with gooks.'

I conceded that he was probably quite right as we had the whole enemy force between us and any element of the 1st Brigade. A chance siting of two people, so far from elements of the 1st Brigade by the MRF boats, could have led to us being mistaken for enemy with undesirable results. Even if we had been recognised, it would have brought considerable attention from the VC. I felt isolated standing on the bank so I slipped in among the reeds myself. The water actually felt warm for a while but started to get very cold as it took away our body heat. Due to the thickness of the reeds, it was easy to rest against them and use them as a prop.

'Colonel, we cannot stay here indefinitely. What should we do?' I said.

He replied, 'Pray! Unless Colonel Emerson saw us go down, it may be some time before we are missed as I am not due back at HQ until late tomorrow.' He added, 'We should sit tight for a while and see what happens next.'

I felt quite secure among the reeds but got steadily colder and, as the sun went down, the *cong moui* (mosquitoes) were insufferable. At times I unzipped my flight-suit and pulled it over my head to get some relief from the insect attacks.

When darkness had fallen, we could hear and see the periodic firefight continuing to the southeast of us and the sky was lit up at times with artillery flares above the contact. At other times there would be total silence except for the constant buzzing of the *cong moui* and my teeth chattering. An hour or two after dark, during a quiet period, I was alarmed to hear Vietnamese voices coming toward us along the canal bank from the direction of the main river. It occurred to me how correct Colonel Archer had been with his evaluation earlier of where the enemy were. Then I became more alarmed as the voices came closer. They were of such a volume that it was obvious that they did not know we were in the vicinity. The VC were not on the canal bank but some 20 metres away from the stream, so they must have been walking along a mounded rice paddy division. There were two of them and they stopped as artillery flares and rounds began to fall back at the main contact area. The colonel and I had climbed out of the water a few minutes before in order to regain some body heat. We were near the top of the canal bank and we could see the two VC silhouetted against the light of the flares. Colonel Archer sat up, did a quick look around, and said to me, 'We can take these two. You take the left one and I will take the right. Aim for the centre of their backs and don't try for a headshot—you might miss. Select single shot', he instructed.

While I was trying to take all this in the colonel knelt up, put his M16 to his shoulder and fired in quick succession, leaving me fumbling to take aim on the left Charlie, who had already started to respond to the Colonel Archer's shot. I squeezed the trigger, not really being confident that I had taken sight correctly, and he was knocked flying. The colonel took another quick look around and said, 'Good work, Son—let's get back into the water.'

We descended back into the foul, cold water, slipped among the reeds, and waited. Our actions did not bring any reaction from the enemy so the colonel was once again correct in appraising that the firefight nearby would mask our gunshots. We observed VC against the light of the flares and heard them talking and moving about many times throughout the night. At times the colonel thought it was prudent to fire and at other times, when I would have fired, he did not recommend it. I had to admire his appraisal of each situation, because we remained undetected by the VC even though we must have dispatched at least ten of them before running out of ammunition. On reflection, I now believe that our position must have been to the rear of the furthermost enemy from the actual contact area. Had we had any enemy to the north of the stream behind us, they would have almost certainly have seen or heard us. Either way we would have been dead.

After running out of ammunition, we decided we should move west along the stream to get away from the area where the dead VC laid. Although we never fired on the enemy when they were in the same area as any previous enemy we had fired upon, it amazed me that their patrols did not stumble across any of their dead. However, small groups seldom moved along the paths so we dropped them in the open. They probably fell among the grass where their comrades would have to walk right on top of their bodies to discover them.

I do not recall looking at my watch once during the night. Time was low on our priorities. As dawn approached we noticed almost a complete cessation of firing from the contact area to our southeast. We did see and hear enemy movement in that area, suggesting that elements of the enemy had swam the Rach Cac River and were escaping to the west. We also observed VC moving along the stream bank not more than five metres from our hiding place. I looked at the colonel for reassurance, indicating that I was not thrilled with this development. He whispered, 'This is good, Son. If we wait for another hour or so, the gooks will have vacated this area to the west and we can start making our way east to the river.'

Just as the sun was breaking the horizon we heard the distinctive sounds of a Huey low down to our east, followed by the sounds of friendly guns. It was a Huey Gunship flying at 50 feet (15 metres) and firing into the vegetation along the canal banks of our stream. The VC went low and did not fire on the Gunship for fear of giving away their position. It was educational to see the enemy response to the Huey's action from the enemy perspective. As they were not in the vegetation alongside the canal, but 20 metres from it, they simply went to ground. The temptation for us was too great with friendlies so close, so we broke cover and waved madly at the Gunship when the chopper was no more than 50 feet (15 metres) from us. The sight of two round-eyed persons, waist deep in the stream directly ahead of him must have given the pilot quite a start. We stood every chance of being mistaken for VC but the enemy would not have reacted in the manner we did and would have more likely fired straight into the cockpit of this juicy target.

The pilot spotted us and pulled his chopper around to assist. The colonel and I started to scramble out of the stream only to see the Gunship taking enemy fire. When the pilot altered his flight pattern and stopped firing, the enemy decided to have a go at it as there appeared to be no other American air support around. The gunners on the chopper started returning fire and our hearts sank as it moved slowly away from us. The gunship crew had no way of knowing that they were actually moving in the direction of more intense enemy activity to the west. Had we had contact with them we would have suggested a retreat to the east. By this time Colonel Archer and I were standing on the south bank of the stream fully exposed but the only enemy we could readily see were southwest of us. Although we did not have time to look too closely, there appeared to be no enemy to our east so they must have all passed our position in the dark.

Without speaking, we both started running in the direction of the Gunship along the raised canal bank. The VC was so intent on firing at the chopper they did not see us behind them. The Gunship had moved behind a thicket of bamboo about 100 metres from us so the VC had to move in order to keep the chopper in sight. The colonel had discarded his empty M16 when we left the stream but I, being trained in a different environment, still had my useless AR15, probably instinctively acknowledging that in our system I would have to account for it somewhere down the line. We still had our respective sidearms but I never thought of using my .38 due to the inherent inaccuracy of firing such a small arm. We were trained to save the revolver to use on ourselves as a last-ditch method of preventing becoming a POW.

We ran along the bank and suddenly came up behind two VC, with their backs to us, firing at the Gunship. The colonel was trailing me as he was having trouble with his back. Luckily the VC to the left side of the bank did not turn around and, holding my AR15 by the barrel, I hit him with full force on the back of the head. He just crumpled to the ground. The VC to the right of the path was alerted and started to turn toward us. Still holding my AR15 by the barrel with both hands, I did a backhanded swipe across his face, then brought the weapon down on his head like swinging an axe from overhead. He also crumpled to the ground.

I will always remember the startled look of horror in his eyes and the mess my second swing made of his skull. As I was swinging backhanded with a twisting motion, I painfully ripped my back and almost collapsed as well.

I could now see VC running diagonally away from us, around the south side of the bamboo thicket, in order to get a better sight on the Gunship. There appeared to be no one between the Huey and us so I started to run with difficulty along the stream bank to the north side of the bamboo, towards the chopper. I looked around and saw the colonel on his hands and knees. Once again I was torn between self-preservation and doing the right thing. The right thing prevailed and I run back the fifty odd metres and assisted the colonel to his feet, thinking that he had been hit. As it turned out his injured back would not carry him any further.

Still loyally hanging onto my rather wrecked AR 15 I stumbled along with Colonel Archer's left arm around my shoulder and my right arm around his waist. I believe the Huey pilot purposefully used the tactic of moving to the south side of the bamboo thicket to lure the VC in that direction. The tactic worked. He then moved to the north end of the thicket towards us where we had a relatively clear line to rescue. However, by the time the chopper had backed up to us, they were starting to receive heavier gunfire, which I detected from muzzle flashes rather than the sound of gunfire. The door-gunners were quite busy firing above our heads. The colonel and I virtually collided with the chopper and, ignoring my injured back, I had little trouble in propelling Colonel Archer into the aircraft with some assistance from a crew member. Another crew member grabbed my arm as I made a death grip on the doorsill. The pilot pulled the helicopter up and away with me straddling the landing skid. Who said you had to be strapped in for take-off!

I did not have the strength to climb into the Huey so I remained straddled across the skid, held firmly by a crew member for the short flight back to the 1st Brigade's forward operations area, where we had started from the previous afternoon. Almost as quickly as the firing started, it stopped. There was just the sound of the Huey's engine. It was almost like closing a door to the noise outside.

A lot of shouting and activity took place after our rescue, but it was all a blur to me. I did not fully comprehend what was going on. For the previous 12 hours or so I had been operating on physical and mental reserves and now I felt totally wrung out. On touchdown, I was pulled into the aircraft. A medic jumped in with us, and we were airborne again, heading west. The medic looked over our wounds, which would have been difficult because we were covered in mud. Colonel Archer insisted that I be looked at first as I had lost quite a lot of blood. The medic had a very friendly attitude and tried to engage us in conversation saying, 'You've had a pretty rough time, Sir!' The colonel must have taken this as a statement and said nothing apart from glancing momentarily at the medic.

I answered, 'Yes!' but the noise of the Huey and the effort in being required to shout to be heard made my head hurt like hell, so I fell into silence as well. The medic gave us both injections in the arm but I have no idea what he administered, possibly a sedative. Twenty minutes later we were touching down at the Tan An Army 3rd Surgical Team helicopter pad, where I had seen so many extremely badly wounded and dead American soldiers unloaded in the past. Even before the engine was shut down, there were medical orderlies at the door with stretchers. Colonel Archer declined the use of the stretcher and, as much as I wanted to lie down, I bravely said, 'No thanks'. The colonel and I tripped along like two drunks towards the hospital.

It was only a 100-metres walk but today it seemed more like 10 kilometres. Half-way there I stopped and said to the colonel, 'We forgot to thank the Huey crew.' I turned around to return to the helicopter pad only to see that the Huey had already lifted off.

Colonel Archer said, 'I've already done that.' I had not heard him speak to the crew.

The hospital was not a permanent structure but a large khaki tent, as were most of the Army buildings away from permanent sites. The reception area, as with the whole hospital, was floored with large sheets of khaki plywood. There was a clerk sitting behind a small khaki desk on a khaki stool and his only function seemed to be to record the names of those being admitted to hospital on a list for receipt of the Purple Heart medal. The Purple Heart is an American medal awarded to those killed and wounded in the US forces. The fact that I was wearing a different flight suit, spoke with a funny accent and had a very unusual ID number didn't worry him at all. Colonel Archer did not want his name on the list and said, 'That is one medal I don't want.' I think the clerk listed his name anyway.

Although the reception area was busy with a number of seriously wounded soldiers being admitted on stretchers, we did not have to wait. It was all ultra-efficient and I was impressed. We were ushered into separate cubicles formed by khaki canvas sheets hanging from the khaki ceiling. As I expected to see Colonel Archer again, I said nothing to him as he was led off in a different direction to me. I lay down on a high, khaki canvas stretcher. The amount of bloodstains on the plywood floor suggested that this was used as a procedures' table. It may have been fatigue, or it may have been what the medic had injected us with, but I went into an instant deep sleep.

How long I was asleep I do not remember but I was aware of someone shaking my shoulder asking if I was awake. I said very groggily, 'I am now!'

'How are we today?' came the voice.

I thought to myself that this was a rather silly question as I was covered in mud and blood, looking very haggard with two days' growth of beard—generally looking far from a picture of health. My new acquaintance was an Army captain doctor, who had a mess of blood on his smock. He looked very tired but was extremely pleasant. He got me to sit up while he took my pulse, examined my eyes and ears and inspected the wound to my head. He informed me that he would have the back of my head stitched followed by an X-ray of my back. He then wished me well and left. Once again I lay down and dozed off, loosing all track of time.

While I was slumbering I was conscious of a male voice and

someone rolling me on my right side. The stretcher was so narrow that a lot of pushing and shoving took place to keep me on the litter. I then felt injections being administered to the back of my head followed by some dull tugging as stitches were inserted. Finally a male voice said, 'We are going to take you for X-ray now, Sir.' I dozed off again but was vaguely conscious of being carried on the stretcher and of the X-ray process. Finally I was taken back to another khaki room. I reflected that they must have lots of these damn khaki rooms. Spike Milligan, the English comic from *The Goon Show*, always said that the reason for so much khaki was so you would know it was the army! Although I slumbered, my mind kept racing until finally I was wide awake and could lay there no longer.

Unsteadily I stood up, slipped on my soiled flight suit and on wonky legs took myself out to reception. I was surprised to see the sun setting through the tent flap. I asked the orderly if I could leave and he said, as he went into another area, 'Just a minute, Sir'. Shortly after the orderly returned with a doctor of Major or Lieutenant Colonel rank who, smiling, extended his hand to me. I didn't even hear his name as my mind was still racing. He said, 'Colonel Archer has left but asked me to express his deepest gratitude and said he would be in touch with you.'

He continued, 'Congratulations!' and pinned a Purple Heart medal on my flight suit.

I responded, 'I didn't think getting knocked around was anything to be congratulated for!'

The doctor laughed and said, 'No, I suppose not, but my congratulations are for what you have done. You Aussies are one of a kind!'

I took that as a compliment. I then asked about getting back to Dong Tam. He made a phone call and announced that there was a chopper going that way shortly and then walked me down to the chopper pad. We talked pleasantries on the way but I really wasn't hearing much as my mind was focused on the events of the past two days and thoughts of how I could have done things differently.

On arrival at the chopper pad there was a Huey Slick with the engine running. The doctor helped me aboard and shouted to the pilot, 'Look after him, he is still rather delicate.' I was covered in mud so the new bandage around my heard must have stood out.

'Roger that, Doc,' came the reply.

'Check yourself in with the Medics at Dong Tam asap,' were the doctor's parting words to me.

Even though the flight to Dong Tam was short, the noise of the chopper engine really made my head ache, so I avoided conversation with the crew. On arrival I noticed that there were no USAF O-1s in their revetments, which suggested they were all out on missions. Sure enough, on arriving at the FAC hooch, I found no-one at home. The first thing I noticed was my lucky charm laying on my bedside locker, right where one of the hooch girls had left it. In my haste to depart on this last mission, I had forgotten to hook it to my flight suit. No wonder I had had so many misfortunes. After showering I was overcome with loneliness in the empty hooch but I just lay on my bunk, staring into the dark with my ears still ringing.

About midnight Major Nelson came in, having been at a 9th Infantry Division briefing. He saw me lying on my bunk and, on seeing the bandage, asked what had happened. He had overheard at briefing of the incident but did not realise that I was the other officer involved. We discussed what I had experienced in detail for an hour or so. Major Nelson said that I should not fly in my condition despite my protesting that I would be OK in a couple of days. He wanted to report the matter to RAAF Headquarters in Saigon and have me medivaced to Australia. I did not want to depart Vietnam under these circumstances as I felt I would be departing a defeated person and preferred leaving on my own terms. After a lot of pleading, Major Nelson agreed not to make a report to RAAF Headquarters but insisted on me taking R & R immediately to justify me not being on the flying roster. Having not had a day off in four months, I agreed. Major Nelson also said something that mystified me at the time but became clear some years later. He said, 'In any case, it is usual for anyone receiving the

"Big One" to rotate back home and be relieved of combat duties.' He was referring to the Medal of Honor I later realised. As I was a junior officer, I had no bargaining power so I have to say that my respect for Major Nelson increased immensely.

After my discussions with Major Nelson I felt a little more relaxed, so I climbed into my bunk and only stirred when I heard other pilots come in about 3 a.m. At 6 a.m. I packed a few belongings and, as I could only see sleeping bodies, I snuck out and made my way down to the airfield. I did not have to wait long to find a Caribou heading to Saigon where, with a small piece of paper in hand that gave me authority to do what I was doing, I headed for the R & R Centre.

Colonel Archer had been a veteran of the Second World War and Korea. A week after this event he received the Bronze Star for Valour when he intercepted and killed a VC by himself. The following week he was relieved of his command due to the back injuries he had suffered on 18 August. In June 2000 I was visiting Robert Archer's widow in Florida. One of his daughters referred to Robert's uncanny resemblance to John Wayne. She then mentioned that if the VC had realised that they had John Wayne and 'Gary' Cooper holed up in 1968, they would have been quivering in their boots!

It is of interest that my action on that day was on the second anniversary of the Australian Army Delta Company's Battle of Long Tan, almost to the hour.

10

R AND R

As my R & R had come quite unexpectedly, I had no real thoughts at first about it. Once it was reality, when I was on board the airliner, I became quite excited. The R & R Centre was immediately adjacent to the northeast corner of the airport terminal at Tan Son Nhut Airport in Saigon. I can't say what the area was originally used for, but it had four-metre-high brick and mud-plaster walls all around it. In 1993 the same enclosure still existed with nothing inside the enclosure but concrete slab floors. It was on these slabs that crude side-less structures had been erected to keep the sun off would-be R & R personnel. Under the stilted roofs were wooden plank beds with no bedding or mosquito nets. There were hundreds of soldiers milling around, passing the time. It reminded me a lot of the Second World War prisoner-of-war films I had seen, with the prisoners exercising in the compound.

This was the first time I had worn my uniform since arriving in Vietnam and it felt rather strange. Among the sea of khaki uniforms was the odd Australian but we were certainly in the minority. An Australian Army Advisory Team (AAAT) warrant officer caught sight of me and he mapped the whole procedure out for me as he had been there for two days. I had no idea about who to see or what to do so he was a great find. Along with an Australian Army corporal, the three of us became inseparable for the whole of the R & R.

The largest crowd centred around three US Army junior officers seated at a table and they were 'Gods'. They had complete authority as to where, when, and if, anyone went on R & R. There did not seem to be much order to the process and one guy told me that he had been there three days trying to get an R & R flight out. The 'Three Australian Musketeers' boldly approached one 'God' at his desk, fearing the worst as we listened to the discontent muttered by those ahead of us. There was a choice of destinations that included Sydney, Bangkok, Singapore, Tokyo, Taiwan and the United States for those going on long R & R. Most of the Americans wanted to go to Sydney but most were disappointed.

Standing before 'God', his scowling face lit up when he saw our uniforms as he himself had just returned from Sydney. Apparently the local female gentry there had provided him with a memorable visit. Thank you, girls! He apologised for not being able to send us to Sydney in the foreseeable future but said he could get the three of us on a flight to Taiwan. We accepted his offer. He had us listed on his papers for travel in two days' time but when he told us he could get us on a flight leaving in two hours there was a near riot from the Americans behind us. We thanked him profusely, quickly grabbed our papers, and left him to be assassinated by the angry mob.

Our flight was a Pan Am Boeing 707 in all economy configuration. As our wheels left the ground there was a loud cheer from the less reserved on board. All eyes followed every move of the lovely flight attendants. Once the plane was cruising the first officer visited the passenger cabin. His name was Don Cooper and he was from Seattle. When he saw my nametag and my pilot's wings he took me up to the cockpit where I spent the entire flight. I was in awe as we approached and touched down at Chiang Kai Shek International Airport. Little did I know at the time that in less than 18 months I would be a civilian pilot with Cathay Pacific Airways, landing at Taipei on an almost daily basis.

When we taxied in, it was early evening and everyone on board was itching to experience the nightlife. The delays in briefing us and allocating hotels annoyed everyone. The main purpose of the briefing was to list all the 'don'ts' we were supposed to obey. 'Don't buy books and music records as they are made without license and will be taken from you before you leave for Vietnam.' 'Do not leave the immediate city area of Taipei.' 'You are an ambassador here; respect the locals and their customs.' 'Do not poke the local girls or you will catch something and your dick will drop off.' With these profound words in our minds we boarded our respective buses for our respective hotels where we collected a handsome allowance in US dollars from the cashier.

My hotel was not too salubrious, but comfortable and not far from the 'O Club' and base exchange. I was so tired that I threw my meagre belongings on the floor, undressed and went to bed in my windowless cell. The noise in the corridors did not disturb me but the banging on my door did. When I opened the door there was a relatively attractive Taiwanese girl standing there. Apparently the hotel had provided girls for each room and the ruckus in the corridor was all the troops swapping girls to their better taste. The one allotted to me started with all sorts of demands and would not go away so I let her in, telling her that all I wanted to do was sleep but she could watch TV with the sound off. For an hour I put up with her demanding money, cigarettes, perfume and all the other gifts she was used to the Americans showering her with. She then started insulting me saying, 'You cheap, Charlie. You no fun.'

I opened the door and pushed her into the corridor. She continued banging on my door with American voices shouting at her to shut up. After putting up with this for a short while, I decided she was probably my responsibility and I should escort her from the building. When I opened the door there was a giant, stark-naked African–American physically throwing her down the staircase. Problem solved!

Next there was another banging on my door again. It was the hotel manager this time chastising me for not paying the girl. I believe I made my point quite clear by shouting at him, 'Fuck off' and slamming the door in his face. It required a second bang of the door as his fingers prevented it from closing the first time. There was some painful screaming in Chinese followed by running footsteps receding down the corridor.

When I woke up it was six in the evening, which confused the hell out of me because I had not got to bed until 9 p.m. My foggy brain was asking how I could be waking up before I had gone to bed! I was so exhausted, that I had slept for 21 hours straight! This would be the last 'full' night's sleep I would ever have. Soon after this I started reliving the events of the night of 18/19 August, most nights, for the rest of my life. There were sometimes variations to the basic events in my dreams, but in general I would toss and turn all night long, thrashing out at the invisible enemy. I would always wake up with a start. I would punch or kick out many times at the pursuing enemy. My wife was often within range unfortunately. My subconscious would tell me there was someone nearby and I would assume I had to go into a defensive mode. This would coincide with me waking up and I would never escape.

After tracking down my two Australian compatriots we headed off for a meal at the O Club. The WO had no trouble getting into the club but the corporal posed a bit of a problem. We eventually removed his corporal strips and I loaned him a pair of my Flight Lieutenant epaulettes. The Air Force epaulettes on an Army uniform went undetected. The standard joke between us then was that we had created a new military rank of, 'Corporal' Flight Lieutenant! We had a three-course lobster meal for two dollars and drank all night for less than ten dollars each. Those US service facilities were fantastic value. The whole time we were in Taipei we seemed to be awake all night and asleep all day, getting over our previous night's activities.

Although we were not supposed to leave Taipei city, most people did. One day we hired a car and went up to the mountain resort area of Peit'ou. This resort is in a volcanic area and all the hotels are piped with volcanically heated water from geysers. The whole place stunk of sulfur but the pine forests and mountain scenery was fantastic. We checked into the hotel for the day and had a delightful meal. As was the custom, a group of girls were marched in and, being the officer, I was given first choice. My choice proved to be a delightfully friendly girl who gave me a sulfuric bath, massage and walked all over my back Japanese-style. Remembering the briefing about one's dick dropping off, I dared not take it any further.

One morning back in Taipei there was a knock on my door and I was met by two SPs (service police). This gave me quite a start and I was quickly formulating reasons why a corporal had been in the O Club in my presence, the hotel manager had sustained a broken finger, and a local girl had been thrown down the staircase. But they were there to escort me to the US military hospital behind the Madam Chiang Kai Shek Hotel to have my stitches removed. I marvelled at how incredibly efficient the system was for that requirement to follow me to Taipei. As it turned out, the medics considered it too early to remove the stitches.

All too soon our five days were up and we were being herded back to the airport to return to Vietnam. I say herded as there were quite a number who had to be rounded up and escorted to the airport physically. There were some very emotional Taiwanese girls wailing for their newfound lovers not to leave. These girls were probably not acting, as the Americans really lavished them in a way they had not experienced before. I recall one guy being carried through the airport terminal by SPs screaming out, 'I 'aint going back to that shit, man!' I don't think he had a choice really.

The flight was delayed as we waited on one AWOL. Further near riots took place as customs officers confiscated books and records from everyone. I had bought a couple of philosophical books by Arthur Conan Doyle for Don Washburn but they were not confiscated for some reason. On the flight back to Saigon I don't think I have ever seen such a large group of totally sombre and dejected passengers. Personally I thought that R & R was non-productive. After experiencing 'life' for a few brief days, it was hard to face the realities of war again. I certainly did not feel refreshed nor did I want to get stuck back into it, particularly after my experiences with my government's attitude in Vietnam.

The evenings in Taipei had been quite cool. In contrast, as soon as the doors of the airliner opened at Tan Son Nhut we were hit by a hot, humid blast of air with the aroma we had all grown to dislike. The lovely flight attendant wished each and every one of us a safe tour in Vietnam and her tone came across with complete sincerity and concern. But, she went basically unnoticed as each of us walked past, deep in thought, and confused about what was to follow.

It was mid-evening and I did not feel much like travelling back to Dong Tam, so I found a place serving meals and wandered off to the R. & R. Centre where I knew I could lay down for a while without being considered a vagrant. There was a US Army PFC (private first class), who I recognised from Dong Tam waiting for an outbound R & R flight. I had seen him playing a trumpet one day during a 9th Infantry band practise. He was wearing a Bronze Star, Army Commendation and Purple Heart among the usual campaign ribbons. Being a non-combatant he must have bought the ribbons at the base exchange. The giveaway was that he was wearing them back to front with the Bronze Star bottom left.

There were other eating and sleeping facilities available but it would have taken me an hour or so to get access to them. About three in the morning I was convinced there was no way that I was going to get any sleep so I went down to where the transport aircraft loaded up. It wasn't until mid-afternoon that I found a Caribou heading in the right direction. It was quite a milk-run but after three or four stops I recognised Dong Tam and got off. The ease with which I wandered around from place to place had me convinced that I could have spent my whole tour in Vietnam drifting about and no one would have been any the wiser.

When I got back to the FAC hooch no one was there. I felt quite lonely. The hooch girls fussed over me a little when I gave them some small trinkets from Taiwan. The gratitude on their faces was rewarding. They disappeared for a while and came back with a cake they had made for me. Neatly in the centre of the cake was a fly, which I did not say anything about.

Over the next couple of days all the FACs drifted through but none of them were all that interested in my tales of R & R. I felt as though the R & R was a distant memory. On 27 August I had one pre-planned strike at WS 996 545 in the western part of our AO. After placing in the air strike, out of curiosity I flew to the area of the contact on 18 August. The Raven was still there and the general area showed signs of recent war. I could see no evidence of human occupation. All was quiet, almost serene, and it was hard to accept what had taken place there only ten days previously—death, mayhem and destruction.

On 29 August we received a new pilot, Captain Larry Mink. He was an old friend of Don Washburn from their West Point days. They were of similar character and got on famously together. Larry, like most of the other Americans I trained, had a little trouble with the propeller and tail-wheel configuration but took to the combat flying very well.

At the end of the month Major Nelson decided to have another go at RAAF HQ. After we had put in our quota of air strikes for the day, Dick Nelson had me fly him to Tan Son Nhut. There he visited 7th Air Force HQ to see why the recommendations for awards on me were disappearing. He discovered, as mentioned earlier, that the Australians had requested that all recommendations on their troops be channelled through HQAFV first. HQAFV would in turn process an Australian equivalent before returning the documents to the US system for American processing. Or that is what HQAFV would have had everyone believe. We now know that the motive of HQAFV was to intercept the US documentation and make it disappear. The Americans in good faith, handed over their recommendations and did not keep copies, as they thought the Australians would return them. Essentially there was no change to the situation except that the Americans were now aware of the actions of the Australian Government. With this knowledge, there was a period where the Americans processed awards to Australians without reference to

HQAFV. However, these processed awards were intercepted by the Australian Government and placed on file in Canberra without notifying the recipient.

By 1 September I had forgotten about the stitches in the back of my head, and the area had become infected due to the unsanitary water we washed in. Dick Nelson had me fly him up to 7th Air Force HQ to tackle the issue of my awards with the Australians once again. While he was busy with HQ I sought out a medical facility near the airport where I had my stitches removed and was given a penicillin shot to help clear up the infection. Major Nelson was apparently sent from office to office but received conflicting and misleading information regarding the awards. He did not come away with any clear idea about Australian policy.

Not long after we arrived back at Dong Tam a couple of us received word that we should report to 9th Infantry Division HQ to receive Vietnamese awards. General Nguyen Viet Thanh officiated—he was to die later in the war when his helicopter was shot down. I received the Cross of Gallantry with Silver Star and the Honour Medal First Class. The Vietnamese insisted on sending the paperwork to HQAFV and again it disappeared. On a flight to Saigon in 1971, when I was working for Cathay Pacific Airways, the company load clerk visited the cockpit with the flight documentation. He asked if I was the Cooper who had served in Vietnam in 1968. On confirming this, he handed me the certificate for the Cross of Gallantry. How he ended up with it, I never found out.

The Vietnamese parade gave Major Nelson an idea and he had the army present me with some of my US awards. They were presented then taken back with the promise that they would be made permanent as soon as approval was received from the Australian Government. This never happened.

The day was nicely completed with a barbecue; the steaks were supplied by the crew chief. He just happened to join in the chain of men unloading the aircraft when the supply Caribou was being unloaded. He put a crate of steaks on his shoulder, walked around the back of the unloaded cargo and disappeared into our hooch. What an enterprising individual he proved to be.

A couple of days later I took Larry Mink up to Bien Hoa to pick up his aircraft from heavy maintenance. The controls had been binding badly and the longitudinal trim was out considerably. Under the floor the maintenance crew found over 20 pounds of spent M16 shells. These had progressively built up over a period of time as a result of firing the weapons out of the side windows of the aircraft. As the shells ejected from the chamber, not all of them fell outside the aircraft. If you were manoeuvering, negative 'g' could throw these shells up into the controls, jamming them with catastrophic results. I understand this was the cause of the demise of a number of FACs and Army pilots. Generally we would fire out the left side of the aircraft. Being right handed it was easier resting the AR15 across my chest which would also steady the weapon. I made the mistake once of firing out the right side and got a face full of hot shells as they are ejected from the chamber on the right side of the weapon.

While I was sitting in the waiting area at maintenance a man came in dressed in faded jeans and a bright Hawaiian shirt. He had bleached fair hair and looked as though he may have just come from a surfing beach. There wasn't a hint of military about him. He was quite pleasant and we spoke for a while but he did not reveal much about himself. He did introduce himself as Sam Deichelman and I recall him saying how he had just met his brother in Saigon, who was serving in the USAF. Sam eventually shook hands, and said goodbye. He climbed into a freshly overhauled and unmarked O-1 and took off. My curiosity about this intriguing person caused me to ask the crew chief about him. He just said, 'CIA. We have been asked to give him an aircraft, no questions asked.'

From 7 September I spent nearly a week on ALO duties, not flying the O-1 at all. On the first day, just before 5 a.m., I made my way down to the chopper pad at Dong Tam as directed by Major Nelson the night before. As usual, my briefing was perfunctory: 'Report to Colonel Hemphill at the chopper pad at 5 a.m. and wear full grunt-gear.' This sortie was to be a five-day search-and-destroy operation by the 3rd Brigade, 9th Infantry Division, just to the southeast of Saigon. On these operations anyone wearing something unusual or bright, like a flight suit, would be a prime sniper target.

Remembering the 'thorough' briefing from the night before, I drifted around the chopper pad looking for Colonel Hemphill but I could not find him. On sighting a marshaller I asked him, 'Pardon me, but do you know where Colonel Hemphill is?'

'Uh?' he responded.

'Do you know where I can find Colonel Hemphill?'

'Who the hell are you?' came the reply.

'I am the ALO,' I pronounced. To the American ear this would have sounded something like, 'I am the Oil O!' And there I was, feeling very conspicuous in my brand new grunt-gear, standing out from the seasoned combatants, displaying captain's bars while emitting an Australian accent.

So, I received another, 'Uh?'

I then decided to spell it out, 'air liaison officer', while at the same time pointing my finger at myself. With an element of doubt, and a sideways look, the marshaller responded, 'Shit, Colonel Hemphill left here 30 minutes ago. Look, get on that Chinook.'

Thanking him for his help I climbed aboard the Chinook he indicated and, being the last one on, I was seated at the rear ramp. Next to me was a first lieutenant, whom I took to be the commander of the platoon of men already onboard. After a few minutes of silence I said, 'Good morning', to the lieutenant who responded by saying, 'Who the hell are you, Sir?'

So I said, 'I am the Oil O'.

'Uh?' Being a quick learner I then said, 'I am the air liaison officer'. The Lieutenant then inquired, 'What the hell are you doing on this bird, Sir?'

I imagine he thought that he was about to be replaced by a greenhorn captain. Greenhorn, yes, replacement, no! I advised him that I had been told to take this flight by which time we were getting airborne and the conversation ceased.

It was a hot insertion! Last on, first off! But as the Chinook ramp opened at the destination the whole platoon raced out over me, making me think that either the pay officer was out there somewhere or it was an R & R stop. I was getting in everyone's way, and was the last off. I had one foot on the ground and the other still on the ramp as the Chinook started to move forward, leaving me feeling somewhat isolated. All my recent acquaintances were now lying on the ground in threatening poses behind rice paddy mounds, so I did the same. Spasmodic small arms fire started to be exchanged, I then thought to mention to the lieutenant that, 'I don't think I should be here!'

The lieutenant said something to the effect of, 'Right on', and, as he had already established that I was no threat to his command, he dropped the 'Sir'.

I soon spied a lieutenant colonel with a radio operator nearby so I crawled over to them, and said to the lieutenant colonel, 'Sir, I am not supposed to be here.'

'Who the hell are you?' he replied.

'I am the Oil O, Sir', I said.

'Uh?'

'The air liaison officer.'

'Christ, what are you doing here? The colonel has been looking for you.'

I did not have an intelligent answer to this. Fortunately the lieutenant colonel continued, 'I'll get the colonel on the radio.'

After a short exchange the lieutenant colonel advised that Colonel Hemphill was on his way. In what seemed like days to me but was only minutes, the command and control OH23 Raven landed right next to us and I jumped on board. As we lifted off, and with a half smile on his face, the colonel shouted, 'Where have you been, Cooper?' I was pleased that he did not inquire, 'Who the hell are you?'

Not wanting to elaborate I merely said that I would tell him over a beer some time. By this time it was about 6:30 a.m. and, having already been in contact with the enemy, I knew it was going to be a long day. Nevertheless, I was thankful to be above the contact, where I belonged, and not in it as I had just been. One day I would like to meet that marshaller who put me on the Chinook.

A forward base camp had been set up on 'VC Island' between Cat Lai and Phu Thanh, eleven klicks southeast of Saigon. By late afternoon I was feeling particularly weary and asked where I was to sleep, as the only structure I could see was a sandbagged tent. This was the command HQ where I had to man the radios while not flying. The colonel, with a smile on his face, said, 'The BOQ is just over there.' He pointed in the direction of a number of grunts digging what to me looked like graves. They were actually digging-in for the night. In my frazzled state I could not face digging a fox-hole, so I threw my groundsheet down behind the sandbags and tried to get some uncomfortable sleep with the constant buzzing of mosquitoes in my ears. At about 2 a.m. VC mortars peppered the base and I wished I had had the energy to dig a foxhole.

At dawn on the second day we were airborne again for a full day in the command chopper. Each time I flew it was with a different company commander, ranking anywhere from lieutenant to 'bird' colonel. By the end of the day, my blind terror of the operation had given way to a resignation—I was going to die in this undesirable country in undignified circumstances. During the day I had noticed a few tents and other living facilities progressively appearing at the base camp. Some of the foxholes had been abandoned so I took up residence for the night in one of those. Unfortunately it rained on and off for a good part of the night. Each time it rained, my foxhole filled up a little bit more with water until eventually I had to try and sleep in a squatting position to keep my important anatomy from saturation. Anyone who was not infantry tended to be pooled together. Being Air Force, I was paired off with the padre, Tom Carrol, who had recently replaced Chaplin James D. Johnson. I found him a little naive as he had not yet modified his reverend ways to suit the combat lifestyle. This was his first time in the field and he was not going to sleep in a foxhole. I guess he figured the Lord would protect him. In the morning, after another night of numerous incoming mortars, I noticed he was sleeping in the foxhole next to mine.

The constant search activities of the grunts pacified the surrounding country so we were getting less sustained night attacks as the week wore on. The forward base, being on an island, enabled security to be established somewhat easily. By the third night the padre and I were set up in a four-man tent with stretchers to sleep on. A large communal ablution site had also been erected. It was a marquee tent with no sides and a wooden pallet floor. Everywhere I went with the Army it was wooden pallets or mud underfoot and khaki canvas or open sky above. I still could not understand why they bothered to put a roof over the shower. What would it matter if it rained while you were having a cold shower? Walls for privacy would have made more sense. One night I was talking with a grunt showering alongside of me when an American female voice joined in the conversation. Turning around, there was the beautiful Sandy, a Doughnut Dolly, who happened to have been walking by and already knew my showering partner. As my blood pressure raced to the lower part of my anatomy, I quickly turned away and continued the three-way conversation over my shoulder. I hope she did not consider me rude with my back to her, but facing her would have been much ruder.

During the third day we took a couple of small arms hits on the command chopper. A major had a round go through his thigh and he was squirting blood from an artery wound. We took a short recess while we got him to the medics and found a replacement. I did feel a little ashamed at my almost happy feelings when the major was shot as my mind told me we could go home for the day. This was not the case—it merely prolonged the day.

Even though there was gunfire nearby, Vietnamese children kept following us about trying to sell Coca Cola, chewing gum, condoms, cigarettes and matches. Most of it was stolen from the Americans in the first place, I supposed. They were also selling their big sisters or little brothers at a dollar a go. There were no adult Vietnamese around, only the children. One of these young salesmen called himself '*Di Di Mau*'. In English this basically meant 'piss-off'. As everyone had told him to *di di mau*, he figured that it would be a good name. He was a bubbly, friendly seven-year-old who I understood to be an orphan living just by his wits. He would follow us around the base, into our tents, shower or toilet, babbling away in broken English. The fact that few people bought anything from him did not seem to deter him at all.

After a nightly mortar attack on the base, there would sometimes be unexploded ordnance lying around the next morning. Until Army ordnance disposed of it, you merely avoided stepping on the duds and did not wander around in the dark without a flashlight. One morning there were about four ARVN troops doing trade with two whores in an old bunker. Another ARVN threw an unexploded rifle grenade in through the entrance, just out of mischief. The grenade exploded inside, killing all six. What a way to go! The padre attended the mess and was terribly shaken by his first experience with violent death in the field.

The unexploded ordnance claimed another victim one night—little Di Di Mau. He must have stepped on an unexploded grenade in the dark. His remains were found near the perimeter of the base, his inventory of merchandise spread among the remains. It was hard to accept that this pitiful pile of mashed flesh had, only the previous day, been a delightful little boy. His remains were not claimed by anyone so he was buried where he was found. The words, 'My name is Di Di Mau. You buy Coke, gum?', has haunted me ever since.

Although the forward base continued for some time, I exchanged places with Andy Anderson on 12 September—none too soon! I was not really enjoying the continuous time in the field. I developed a great admiration and respect for the infantryman from the experience but it was like a homecoming to be back flying '981, even though I only did two boring VR missions.

In the latter half of September I started training two new pilots. One was Major Chuck Cage, a very calm, friendly person who did not use his rank to gain preference. He checked out in the minimum of time. The other pilot was Captain Jo Nuvolini, who had the usual propeller/tail-wheel trouble but otherwise was another fine pilot and officer.

On a flight to Bien Hoa I was talking with the crew chief at maintenance and he referred to Sam Deichelman. He said, 'Do you remember that pilot in the Hawaiian shirt who was here earlier this month? We just had a call asking when we intended to dispatch him with the new O-1. He left here on 6 September and did not arrive at his destination. They have only just missed him.' Apparently Sam was a Raven FAC (they did CIA-type operations) and was flying the new aircraft up to Long Tieng in Laos. He disappeared and has not been found to this day. There was a story going around that he wanted to see a lake called Tonle Sap in Cambodia and it is thought he may have taken a detour via there. The surface of Tonle Sap can be glassy smooth with the clouds reflecting in it like a mirror. He could have misjudged and flown into the glassy surface of the lake. If he did not fly into the lake, one day his remains and his 'new' O-1 may be found rotting in the jungle. Tragically his brother was also killed in Vietnam in a midair collision with another aircraft. Both brothers were the sons of a USAF general.

I did not work with troops-in-contact again until the night of 20 September at XS 397 394, which was only four klicks south of Dong Tam on the other side of the My Tho River. This area was heavily covered with coconut palm and in the flashes of exploding bombs, it was mesmerising to watch the tree trunks being blown into numerous logs and bounce away.

On 25 September I was putting in some pre-planned air strikes near Ap Bac, some forty klicks northwest of Dong Tam, massive area of swamplands and abandoned rice paddies that extends into Cambodia. It was featureless except for the odd isolated tree growing out of what must have been slightly raised ground. Ap Bac fort was near the junction of four canals, all of which was overgrown with reeds as the area had been long abandoned by Vietnamese farmers. Nature was reclaiming her land. Along the side of the Muoi Hai Canal was a long and straight sealed road that led nowhere anymore. As I was directing the fighters I noticed the cloud getting lower and thicker, reducing visibility. The fighters were having some trouble handling the conditions and did a couple of abortive runs. Eventually, one of the F100s said, 'Hey, FAC, we will make this our last pass as I don't like the way the weather is turning.' As the fighters were pulling up to 5000 feet (1500 metres) on top of the low scud, they could see what I could not. Approaching from the east was a monster of a thunderstorm. The wind started to pick up and the sea of reeds below was lying down flat under the great force. I was caught by surprise at the rapidity with which the storm approached. As the fighters left they said, 'Hope you will be OK, FAC. This looks like a bad one.'

How right he was! I started heading back to Dong Tam but a black wall of water and typical rolling clouds blocked my way. Because I couldn't get to Dong Tam I headed southwest towards Vinh Long. Soon I met the same wall of water. I decided to head north for Moc Hoa, a small outpost near the Parrot's Beak but my path there was soon cut as well as the classical 'hook' of the thunderstorm closed around me. My only option was to get on the ground and the abandoned highway along the Muoi Hai Canal would have to be my runway. The highway was into wind, which was blowing at about 50 knots (90 kilometres per hour) plus by this stage. On late finals I could see that the road surface was overgrown with grass and bushes along the edge so I landed with the flaps retracted because with the flaps lowered I would have been flying backwards in the gale-force wind. I touched down smoothly in the three-point attitude with the control stick fully back, expecting to tip over, but the surface did not have any holes and I pulled up in the length of the aircraft.

The stormfront was on me as soon as I landed and the full fury of nature tried to devour '981 and myself. I instinctively felt for my flight-suit zipper and found my lucky charm was in place, which gave me comfort. Silly really! I was so involved with my dilemma that I did not think to radio my intentions to Tamale Control; however, they knew where I had been working and I knew a rescue chopper would be on the way at the earliest opportunity. At first I considered abandoning the aircraft and seeking cover in the canal, then I decided to stick with the aircraft—she had stuck with me all these months. I knew I could always roll out of her if the wind took over and looked like flipping her. For about five minutes I sat in the aircraft holding the park brakes on and basically 'flying' it on the ground. The deluge of rain and thunder was deafening but the storm passed over in five minutes. As the wind died down I watched the monster storm crawl away towards Cambodia. Apart from the receding thunder, all was quiet and still. I started to feel uneasy about what might be out there in the reeds so with great haste I started the motor and rolled down my makeshift runway, pulling '981 into the air early and accelerating just above the half-metre length grass. I called Tamale Control, not letting on that I had landed, and advised them, 'All OK, with you in 20 minutes.'

Troops were in contact with the enemy on 27 September at XS 105 617 near Xom Tre. Joe Nuvolini was in the back seat for familiarisation and I became a little annoyed at him as he was trying to record the radio transmissions on his tape recorder. The recorder was on his knee and when I did some fairly extreme manoeuvers, the recorder kept restricting the movement of the control column. The ground troops were spread out over a large area along the north bank of the My Tho Canal. I could see a group of grunts lying on a bank in an open field. An F100 had released his napalm late, but before I could tell him to stop. The napalm boiled on, spreading toward the hapless grunts. They could see it coming and took flight. With a sigh of relief, the napalm stopped short of them but my heart was in my mouth as I watched this event in seemingly slow motion. As the napalm had been spreading I had involuntarily stomped my foot on a make-believe brake pedal, as you do when you are a passenger in a car. It may have helped! There was a total of nine fighters on this target. Two days later, I was again called out to troops-in-contact in the Xom Tre area and I directed a further eight fighters and one Canberra in support.

The Xom Tre area was becoming a hot spot. On 3 October I was again dragged out of bed at 6 a.m. to assist troops-in-contact this time at XS 082 616. The day before I had been at Bien Hoa, socialising with the fighter pilots. I had only landed back at Dong

Tam at 3 a.m. so three hours of sleep had me feeling in less-than-peak performance. I approached the contact area at 4000 feet (1200 metres) to get a better appraisal of the general layout. About a klick west of the ground forces I spotted another group of soldiers. Checking with the ground commander, he confirmed they were not associated with him. Pulling out my binoculars I could see about 40 VC heading towards the contact site. As I was high up, they could not hear me and they probably figured all the attention would be at the contact point, allowing them to move freely in daylight. I gave the first two pairs of F100s I briefed random headings for their attacks, and put in a marker rocket 200 metres from the VC unit. This had the desired effect of bunching up the enemy as they reversed back up the trail. The first bombs were among them in no time and I noted only about ten of them emerged from the smoke and hide in a canal. As there were gunships in the area, I gave the position to them and they put in further attacks on the canal banks and reeds. I controlled a total of ten fighters on the target and the ground troops gave us 30 KBA by the time we had finished. I was back in bed by midday.

Due to the enemy sightings west of Xom Tre, intelligence arranged a daylong operation in the area on 4 October. There was to be B52 strikes along the Hai Muroi Tam Canal while the FACs controlled fighters bombing along East My Tho Canal. Five FACs were involved in relays to keep up a constant string of air strikes along the main canal where the enemy battalion was thought to be. The B52s were to create a blocking action to the south. This was to be followed by an airborne assault by ground troops. General Ewell and a number of colonels oversaw the operation from the command chopper south of where the B52s were dropping their bombs. All this got underway at 5 a.m. which was unfortunate, as there was still fog in the area when I arrived as the first FAC on target. By the time the first fighters arrived, conditions were not much better.

I had a formation of eight F100s. I had not worked with eight before. There were some breaks in the cloud but in most areas it was still on the ground. I could pick up the canal through breaks in the cloud but could not get underneath. Rather than waste any ordnance and jeopardise the operation, I decided to have the fighters do level bombing in pairs and release in salvo as they flew directly over me. This was rather unorthodox but it was worth a try. I set myself up in a tight orbit at 200 feet (60 metres) over the canal and had the fighters fly 500 feet (150 metres) above me. As I looked up at the belly of the fighters I called, 'Release now!'

The first pass was a little scary as I watched the mass of 750-pound bombs released and sail past me. It worked well, though. While I was orbiting over the hole in the cloud a .30 calibre round went through my engine cowling. I felt for my lucky charm just as another round went through the floor and rear seat. It was Joe Nuvolini's lucky day as he should have been in the back seat but missed the flight because he did not wake in time. He was safely tucked up in bed at the same time as a round with his name on it went through his seat. On one of the bombing runs there was a giant secondary explosion that threw mud into the air, hitting my windscreen.

Major Chuck Cage replaced me as FAC and he was able to get below the clouds that were starting to lift and break up. However, being so low he took three rounds through his aircraft and had a fighter hit by ground fire as well. But the whole operation went well and an enormous enemy underground system was uncovered. The results of the operation were, 139 KBA and 30 sampans, 7 structures and 126 bunkers destroyed, including a tunnel complex. They also captured 3000 grenades, 6 anti-aircraft guns, 8 mortars, 20 machine guns, 160 rockets, 11 000 rounds of ammunition, 50 mines and tons of clothing and food supplies.

Post-flight inspection revealed some minor damage to my propeller from the secondary explosion. Joe Nuvolini was to fly it up to Bien Hoa for repairs with me in formation to fly him back. It was around this time that I made contact on the radio with Mac Cottrell who surprised me by saying, 'Two weeks to go!' He told me that Roger Wilson had just left and he was going home in a few days. As soon as we landed at Bien Hoa, I got on the phone to RAAF HQ to enquire if I had erroneously not received my orders to return to Australia. The voice on the other end of the phone seemed surprised that I wanted to go home but promised to get my orders through asap. I had been overlooked, 'Sorry!'

Due to a series of events, Joe Nuvolini and I ended up spending two days at Bien Hoa. I threw him into the Lion's Pit at the fighter hooch as the FAC representative, then disappeared early to get some rest in my 'secret' abandoned hut. The mechanics played around with the aircraft most of the day and ended up doing an engine change to be on the safe side. That night there was a party which the Doughnut Dollies sponsored. Joe Nuvolini turned up in a terrible state of inebriation but feeling no pain. Even the presence of lovely female company wasn't enough for me, I found I could not enjoy myself so I left early for my abandoned hut.

On the flight back to Dong Tam I flew back seat to Joe while he put in two pre-planned air strikes. He seemed to operate much better hungover than he did sober. The next few days were tied up with administrative flights between Dong Tam, Bien Hoa, Tan Son Nhut and Vung Tau. On one of the flights to Tan Son Nhut I took the opportunity to visit RAAF HQ. The conversation I had had a couple of days earlier with RAAF HQ about my going home had not been recorded and no-one had any knowledge of what was discussed. I was furious and demanded to see the air commodore, who asked me if I had compassionate reasons for returning to Australia so soon. This left me dumbfounded. They must have had me confused with someone who had only just arrived in country that month. Eventually it was agreed that I should plan on returning to Australia at the end of the month. The clerk gave me an envelope of papers, which I took to be my orders, and told me to take them to Vung Tau. I really wondered what the function of RAAF HQ was. At Vung Tau it took over 24 hours to process the documents and I eventually left without a firm return date. In the process they had tied up two combat pilots and one combat aircraft for two days while they satisfied their administrative requirements.

Once again, being non-American, I picked up another oddball job on 9 October. This time it was to address a group of correspondents from *Time/Life*, or some other media, on our interaction with defoliation operations. In the O-1 we were not able to keep up with the C123 Ranch Hands so we would orbit around the centre of their Agent Orange runs in case they picked up ground fire. We could then call in fighters if required. Sometimes we would get a whiff of the evil chemical, which was nauseating. After my address, tea and biscuits were served and I was presented with a Silver Star for the operation on 4 October. It was just a publicity stunt I believe as the paperwork for this decoration disappeared into the Australian system. One of the correspondents was a delightfully demure and retiring blonde in her early twenties, wearing a smart khaki uniform. She certainly caught my eye but I paid her little attention as there appeared to me to be no future in it. After a while I became conscious that when I moved from one group to another, she also moved. We eventually separated from the group into a corner and obviously had a lot in common. I was quite infatuated. We talked for a couple of hours until the correspondent group had to catch a plane back to Saigon. Her name was Heidi Heidemann, an American-born German, and I was sad to see her leave as she provided the first real female company I had experienced in months. Two letters from her followed me through the system, catching up with me a couple of months later. Although the letters were totally innocuous, I dared not encourage any further contact.

On another maintenance flight to Bien Hoa I ended up spending three days as there were problems with aircraft availability. The attitude of my RAAF superiors also had convinced me to give up trying to make an effort so I put my feet up and waited to be summoned. During this time I managed to get two rides in the back seat of F100s, doing ground-attack missions. It was great to get back into a jet again. Major Nelson displayed his humour by sending me a signal saying, 'All is forgiven, please come home!'

Several weeks prior to this I reported the wreckage of an aircraft I had noticed in a small stream. The only recognisable component was a mangled wing, which had been dragged across the stream to form a makeshift footbridge. It was well used as the trails to and from it were quite distinct. As with other wrecks I had seen, I was interested to find out the history and the fate of the crew. I kept asking about it but there was never any return advice. Finally, a ground team went in and established it was a Cessna that had gone missing many months ago. Probing around the submerged wreckage the remains of the two pilots were located.

On 21 October my morning started with a call by Sergeant Cover. 'Sir, wake up! Troops-in-contact near Xom Tre.' Arriving at the revetment, Jake had the engine running but my book of maps was missing. He did not know where they were. Why anyone would steal a bunch of maps I could not fathom. For some time Jake had been complaining about the weight of the maps as he was supposed to put our maps and survival equipment in the aircraft as a part of his preflight tasks. It is hard to believe that anyone could be so irresponsible as to throw them away but under the circumstances I believe Jake did. I located a 1:250 000 scaled map in the Conex, which would have to suffice, and took off heading west. I spent 6 hours 30 minutes in the air that day on two sorties, putting in seven air strikes. On the first sortie I probed around with the bombs for a while about a klick from the friendly troops. They had pulled back and no longer had visual contact with the enemy. When I started to pick up ground fire I knew I was getting close. On one pass my rockets would not launch from the right side and afterwards we found that the ground fire had severed the wiring to that side, although I was not aware that I had been hit at the time. The damage may have been caused on a flight prior to mine and not detected during the pre-flight inspection as it was only the electrical wire cut. In the dive, trying to make the rockets work, I went far too low and could hear the continuous rattling of the distinctive AK47 automatic fire. After expending the left rockets I then had to complete the mission using hand-held smoke canisters thrown out of the window. This meant getting in close and low and virtually lobbing the canister at the feet of the enemy. It was rather scary doing it this way but I found, being down low and approaching from different headings, the enemy could be surprised by the delivery. They had to fire vertically in the air hoping I would fly into their bullets either going up or down.

At one stage I spotted a frail bamboo bridge over a stream that looked like one the enemy might head for during their retreat. I told the fighters I was going to put in a mark on the bank and they were to bomb centre stream as I figured they would have trouble seeing such a small target. I rolled in and to my amazement, my rocket hit the bridge, collapsing it. I told the fighters, 'Hang on, I'll get you another target—I just knocked that one over myself.'

The reply came back, 'Hell, Tamale, you don't need us!' His wingman added, 'Yeah, Tamale, let's know when you've finished down there!'

The next day I was doing a VR mission in the VC Panhandle when an overgrown, box-shaped canal caught my attention. I saw in it what appeared to be several 44-gallon drums. There was no habitation anywhere near here and it was in the free-fire zone, so I called for a fire mission from the nearest FSB. They did not want to expend artillery on such a target, so I decided to put a few smoke markers into the drums. Although I got very close, I did not get a direct hit. While I was flying around at about 200 feet (60 metres) trying to work out what the drums might be for, I heard AK47 automatic weapons fire. I slammed on full power and started a climb out, looking for the source of the ground fire. To my amazement, not far from where I had been circling, I saw a lone VC with a water buffalo. He had his AK47 resting on the back of the buffalo and was firing at me. It must have been a docile beast as it did not move while its owner was firing his weapon just centimetres from the animal's ears. The buffalo was not harnessed with a cart or any agricultural device, which suggested to me that the VC was wearing peasant clothing and using the buffalo to make it look as though he was a farmer.

Of late, I had started to get very short on patience and, forgetting my vow not to take chances anymore, I went into a rage. I pulled out my AR15, selected automatic, positioned the gun on the lower window frame and hurtled the O-1 at my quarry. This was foolish as there could have been many VC down there but logic had failed me. I dove straight at the VC not being able to fire until I was over the buffalo—the angle between the propeller arc and the wing strut was very small so it would have been futile to fire ahead. As I flashed over the buffalo I racked on 90 degrees of bank in a climbing turn and, when I had a full view of the buffalo out of my left window, I opened fire, emptying a magazine into the scenery no more than 50 feet (15 metres) below me. As I climbed away looking over my shoulder, I could see the VC flat on his back and the buffalo buck-jumping as it bounded away. I was then overcome with remorse at what I had just done. My concern was more for the buffalo than the VC. By the time I had reloaded with the intention of finishing off the buffalo, it was laying still in the abandoned rice paddy some hundred metres from the contact. After circling for a few minutes, the stillness of the beast and the spreading blood in the water convinced me it was dead. As far as the VC was concerned that was one who would not kill a GI again.

At a reunion in June 2000, I remember a member of the 5th/60th, Tex Balas, telling me the story of when he shot a buffalo. Tex is a typical lanky Texan who even today dresses in full cowboy gear, right down to the pants with braces. He is a person of great character and a delight to listen to. He said they had been on patrol and were taking a rest when a few buffalo walked by. He leapt to his feet and fired on automatic into the lead buffalo. Everyone had thought he had gone mad and the Lieutenant asked, 'Why in the hell did you do that, Tex?' Tex's laconic drawl replied, 'That is the first God dam six-legged buffalo I've ever seen, lootenant!' Walking on the other side of the buffalo, and hiding behind it, was a well-armed VC, hoping to creep by. Tex had dispatched the VC and the buffalo.

On another occasion Tex was on an APC in support of Fire Support Base 'Jaeger', which was being overrun by enemy soldiers. The base was under the command of the courageous First Lieutenant Lee Alley, who won the DSC but should have received the Medal of Honor. He and a group of men were fighting for their lives—the more I heard from Tex, the more I grew to admire and respect the infantrymen. As Tex's APC tried to enter Jaeger they had to shoot VC climbing onto the APC. Eventually, with the enemy fire being so intense, they had to withdraw. The APC that Tex was crewing was hit by an RPG, which killed all but Tex and one other. Tex was not badly wounded but his compatriot was. During the course of the early morning, Tex fired off over 400 rounds in defence of his wounded comrade. Surrounded by enemy, Tex sought cover behind the destroyed APC. His delirious comrade grabbed Tex by the shirt saying, 'You're not going to leave me are you, Tex?' Tex replied in his usual drawl, 'Well hell, where am I going to go?'

Thanks to Tex, they both survived.

Two weeks after the 4 October contact west of Xom Tre I flew over the area to check if the enemy had returned. Approaching overhead the stench of rotting flesh was overpowering even at 500 feet (150 metres). There had been no effort to clean up and the enemy had not returned to claim their dead. With such carnage on the ground, I think it would be some time before anyone would venture into the area again. After establishing that there were no fresh foot trails through the bombed out areas I made a mental note to keep away until the wild animals had cleaned up.

I continued the usual routine of pre-planned air strikes and VR missions, mainly instructing from the back seat, until 27 October, when I received a verbal notification that my combat tour was at an end. Major Nelson allowed me to take '981 up to Bien Hoa to say goodbye to the fighter pilots and get my clearance from 19 TASS. In the early morning I went down to say goodbye to the Doughnut Dollies, who would not let me in. If I had not been drunk I would have appreciated the reasons why. Their compound was out of bounds to male personnel. However, some very tolerant and friendly girls sat outside and had a few drinks with me. I had the foresight to take Joe Nuvolini with me as I was something of an inebriated mess by 3 a.m. Joe flew us back to Dong Tam—one of my last sorties in '981 was sleeping in the back seat!

There was no completion date set for my end of tour so the next day I put in three pre-planned strikes with two B57s and four

F100s, destroying 14 sampans, 7 structures and 7 bunkers. It was a good finish for my tour of duty, with no return ground fire. As there was still nothing official from RAAF HQ, Major Nelson contacted them and was told to release me—I was to report to Vung Tau the next day. On the morning of 29 October I flew to Vung Tau with my worldly possessions in an overnight bag and Larry Mink in the back seat to return the aircraft. In the vicinity of Go Cong we flew into an artillery barrage and I thought, No, not now! Throwing the aircraft into a tight turn I contacted the artillery net and found that the fire was coming from Navy ships on the My Tho River. Looking south I could see them. The impact of the shells indicated that the line of fire was sliding underneath us. My landing at Vung Tau was heavy and untidy, being the worst I had done in an O-1—it was not until 1996, when I next flew an O-1E Cessna, did I redeem myself on the landing. There was a very strong easterly blowing and the crosswind was outside the limits of the aircraft so I landed on the short cross runway, which looked more like a parking area than a landing strip. As I approached over the blast barriers, a gust of wind caught me and I slammed into the runway. The aircraft lost all energy in the gust and did not bounce, but the touchdown was one I would not care to admit to. I taxied in, shook hands with Larry Mink and handed the aircraft to his care and control. I watched him take off and disappear to the west with mixed emotions. Picking up my overnight bag I headed for the orderly room on the RAAF Base to receive my clearance home.

11

THE STRUGGLE TO GET HOME

AT THE ORDERLY ROOM in Vung Tau I announced, with the slight American accent I had gained, 'I am here for processing back to Australia'. Also, wearing an American flight suit probably added to the confusion about who I was. I became agitated when a discussion ensued among the staff without including me. One clerk was talking with another while looking sideways at me, but at a volume I could not hear. As my agitation was obvious, the volume was increased so that I could hear what they were saying. The orderly room sergeant finally turned to me and said, 'But, Sir, you are supposed to be at Bien Hoa!'

'Surely I can get on a C130 here,' I said.

'No, Sir, you do not understand. We have you listed for instructor duties on the OV10 at Bien Hoa.'

I went into a rage, which obviously startled everyone in the orderly room. The sergeant dashed into the next office and collided with a flight lieutenant administrative officer, who was coming out to see what the shouting was about. After a heated discussion I found I had no option but to board a C123 heading for Bien Hoa. On arrival there I found my favourite abandoned hooch. There was only a bed, mattress and mosquito net but I had the whole place to myself. I was disgruntled and depressed and in no way felt inclined to display an enthusiastic attitude.

The next morning I reported to 19 TASS. Colonel Patrick already had word of what had happened and apologised profusely. We must have spent about two hours just discussing things in general. I believe he was just calming me down and it was good of him to devote so much of his valuable time toward the well-being of a junior officer, who was a foreigner anyway. I considered I had an 'out' as my big question was, how can I be an instructor on the OV10 if I am not checked out on that type of aircraft. Colonel Patrick explained that the OV10 was new in country and there were enough instructors but few with combat experience available. While most of the pilots were upgrading from other FAC aircraft, there were still a few who were checked on type but without combat exposure. I was to ride back seat with new pilots in country to teach them the FACing business.

I must say that I was impressed with the OV10 and its performance. It was a dual control aircraft and most of the pilots let me do much of the flying from the back seat. The aircraft handled nicely, although I found sitting up higher than the front seat man a little uncomfortable initially. Each morning I visited Colonel Patrick for a coffee at his invitation to see if he could do anything about my return home. Many people considered Colonel Patrick a bit of an old woman. I liked him as there was no nonsense or bullshit with his approach. He was honest and could be taken on his word.

Over a period of five days I did nine sorties on the OV10. It was a versatile aircraft. It was interesting to receive clearance for a mission at the same time as the F100s I was going to control, take off before them, and then have the target burning by the time they arrived. I would then be back in the crew lounge reading a magazine before they got back.

On the morning of 3 November Colonel Patrick told me he was successful in getting my orders to return home. The papers would be completed by his staff so there would be no chance of the RAAF screwing things up for me. I could collect the papers after lunch. While waiting I did another back seat ride in the OV10 and on our return trip we did a beat up of the field. The tower had cleared us to do this but requested we keep north of the runway, away from traffic. I approached the airfield at high speed, diving down to 200 feet (60 metres) over the touchdown end of the runway. I pulled up into a vertical climb where I rolled right through 90 degrees, allowing the nose to fall into a 45 degree dive, rolling into a 90-degree bank turn onto downwind. I then pulled the aircraft straight around onto base leg, dropping gear and flaps, and landed. Because I was unfamiliar with the aircraft I considered my handling of it was rather ragged. The first lieutenant in the front seat said, 'When I have your experience on the OV10, I should be able to do the same thing.' I didn't have the heart to tell him that he already had more hours on the OV10 than I did.

After lunch I climbed onto a C123 bound for Vung Tau. I had no baggage, only an overnight bag containing my shaving kit, passport and a few papers. I had mixed feelings about leaving again. I was leaving behind people I had grown to respect and worked well with. I also felt the job was unfinished and was sorry for those who remained to fight this silly war. I think the motivation to leave was caused by the knowledge that my two compatriots, Roger Wilson and Mac Cottrell, had already left. Without that motivation I believe I would have remained at Bien Hoa and hounded Colonel Patrick to get me back on operations. As far as I was concerned, the RAAF no longer existed and if I could have I would have remained with the USAF. I could now understand why soldiers returned for a second and a third combat tour. The challenges are severe, the adrenalin rushes extreme and the results permanent. There is no other life that can compare.

Arriving back at Vung Tau I was told it would take two working days to prepare my clearances. I was like a bull with a sore head and no-one dared approach me. There was no way that I could even consider going to the officers' mess to socialise. During the night the base had a practise ground attack with the sirens sounding. I did not respond and stayed in bed as I had had

THE STRUGGLE TO GET HOME

plenty of practise at running for the bunker for real. An extremely officious SP, wearing polished boots, buckles and webbing, with his uniform pressed into razor-shape seams, burst into my hut ordering me to run to the nearest bunker. He towered 200 centimetres above me in the standing position but I jumped off my bunk and screamed in his face to, 'Get stuffed!' Startled, he backed off and said nothing but put me on report for 'not playing the game'. In the morning Group Captain Hubble confronted me, advising that he could probably get me on the 10 a.m., C130 to Butterworth, Malaysia, but asked me to settle down a bit. Despite my attitude, there were a few who went out of their way to help me.

On the C130 I sat sideways in the canvas bench seating with my feet resting against the lower unit of eight stacked aluminium coffins. Some were obviously occupied by less fortunate soldiers than myself as the crew chief came around a couple of times during the flight to drain off the vent system.

At Butterworth we stayed overnight and I felt out of place wearing my uniform and eating with silverware at a mahogany table. The serenity and cleanliness felt like another world. By now I was feeling a lot more sociable and I did enjoy a few drinks in the bar with old compatriots. Next morning we took off in the C130 for Darwin via Singapore. It was a long, cold and cramped flight with the aluminium boxes dominating the interior. At Darwin another enemy far more fearsome than the Viet Cong confronted us. The Australian Customs. In those days, and perhaps still, a customs officers' value was gauged by how nasty and belligerent he could be to the public. There were about 30 of us on the C130 returning from fighting for our country. These embryonic Gestapo agents spent over an hour processing us. *Playboy* magazines were seized as illegal and pornographic material, and the owners were threatened with jail sentences for smuggling. They searched through everything we had and emptied our suitcases, leaving the contents scattered about the benches. All the 'prisoners' contained themselves admirably under the circumstances—not one customs officer was strangled!

In keeping with my low social status, I walked myself across to the officers' mess. There I ran into an old friend, Squadron Leader 'Nobby' Williams, who arranged a senior officers' quarter for me, which was a welcome gesture. We went out together that evening and had an enjoyable time which lasted until the early hours of the morning. While I was depressed and not much fun, Nobby kept the evening rolling. Nobby was later to do a tour with the Australian Embassy in Saigon and carry out some FAC duties in Vietnam.

The next morning we took off in a C130 bound for Richmond near Sydney, our final destination. Arriving there, having already been 'greeted' by customs in Darwin, there was no-one to meet us. There were about ten relatives of my fellow travellers, but no official 'welcome home'. We simply skipped in and faded into the darkness in different directions. I went to the officer's mess where, unexpectedly, my wife turned up to drive me back to Newcastle. This was my first view of my son Carl as no photographs had ever reached me in Vietnam.

When I arrived back at Williamtown for duty, the base was starting to wind down for Christmas so I was given leave to report for duty early in the New Year. When I returned from leave I had a couple of familiarisation rides in the two-seat Mirage with Flight Lieutenant Pete Smith, who had distinguished himself as a FAC in Vietnam before my tour. After the training was over I was posted back to 76 Squadron as a squadron pilot. All the pilots had progressed while I was away and I rejoined as a junior pilot.

In February it was decided the RAAF needed a FAC school to train pilots before going to Vietnam and two Winjeel training aircraft were acquired from RAAF Laverton for the purpose. Mac Cottrell and I were flown to Laverton, Victoria, by Qantas to pick up the two aircraft. These we would use to operate in a FAC role to familiarise other pilots with working under the control of forward air controllers.

Mac Cottrell and I were assigned to give FAC exposure to the 5th Battalion, Royal Australian Regiment, commanded by Lieutenant Colonel Colin Khan, which was about to go to Vietnam.

On the lead up to this, we attended a briefing at operational command. The briefing officer, although it was his portfolio, knew nothing about forward air controlling. He was a wing-commander navigator wearing Second World War ribbons—he had not seen action since 1945. He handed us a 1944 manual on forward controlling written by the British Army. Although this was inapplicable, it was all he had. Despite our objections, he insisted that this would be our guide in training 5RAR.

The RAAF operational command was obviously not interested in what Mac and I had learned in Vietnam. To go along with this regressive thinking was not only a waste of time, but would also confuse 5RAR when they began operations in Vietnam. Mac and I decided we would display to 5RAR exactly what they could expect in Vietnam. We used American terminology and made it as realistic as possible with all our training and demonstrations. After the exercise, each unit commander had to write a report on his views of the results. I wrote as I saw it. In the final report by operational command about the whole exercise they considered the best report was written by a second lieutenant who said, 'The exercise was very well planned, was well executed and was a resounding success.' Of course, he had not yet been to Vietnam. My three-page report was judged the worst as I was unnecessarily critical of the organisation and the equipment used. Although the OC operational command signed the report, our Second World War navigator no doubt wrote it. It had me wondering why they bothered to send us to Vietnam to gain experience if they were not going to then use the knowledge, and be guided by a Second World War navigator. They did not want to hear from us or use our hard-earned experience.

While this variety of flying was interesting, I found I could not get enthusiastic about the peacetime military any longer, particularly after this slating over the 5RAR report. I was put on a Mirage Ground Attack Course but I became progressively more disgruntled. During the course we did weapons delivery with live munitions on Saltash Range. Our academic score was based largely on our achievements on the range. Unfortunately, I pushed the minimum height limit to get in closer to the target for greater accuracy and was 'fouled' on several occasions. This led to me scoring penalties. Even after losing 50 per cent of my score in penalties I still managed to achieve second place.

In May 1969 I received word that I had been awarded a British Distinguished Flying Cross and subsequently had the award pinned on me, along with Pete Smith, by Sir Roden Cutler VC, at Government House in Sydney. The citation briefly read: 'In recognition of his courage and devotion to duty while serving in Vietnam.' It did not mention an action or a date. All my US awards were far more explanatory and complimentary than this. I was fortunate that my award came through so soon after the service. Others, like Roger Wilson, at the ceremony had waited two years for their recognition to come through. At that time, all FACs were getting DFCs. It was just a matter of doing the job and waiting for a significant action. The formula used by the government for the issuing of awards to fighter pilots in Vietnam was one award for every 300 operational hours flown. By 1970 the backlog of recommendations necessitated the government to steal hours from 1970 to qualify the recipients in 1969. So, what should have been a display of a grateful Government giving reward to its gallant airmen, ended up very much like a lottery draw and an elimination contest. Arthur Sibthorpe gave gallant service as a FAC and was recommended by the Americans for a Bronze Star. He should have received an Imperial DFC but was 'disqualified' because of a minor misdemeanor while off duty. These incidents highlight the degrading manner in which our government handled the awards issue.

About this stage 77 Squadron was re-formed with Mirage ground attack aircraft. I was posted in as 'A' Flight Commander along with four other pilot officer-rank pilots and a temporary CO. We had 21 aircraft and a full year's allocation of ordnance to use in one month. Naturally I rostered the six of us for nothing but low-level exercises, resulting with live firing at the range. The CO didn't like me and would counter many of my decisions. He even ridicules me to this day, or so I hear. One night I set off on a low-level cross-country and I was to radio back if the weather was suitable for the remaining aircraft to follow. There was a little light rain and low cloud around the mountains and coastal regions but, apart from that, I whistled along at 420 knots (770 kilometres per hour) at 500 feet (150 metres), AGL (above ground level) clear of cloud. I radioed back that all was well only to get back to Williamtown to find the place in darkness. The CO had cancelled the following aircraft and stood everyone down. Not only was that embarrassing, it was a shame as, up to that point, all the commanding officers I had had, both Australian and American, had been first-class and competent.

By July I had become disillusioned with the RAAF and resigned my commission, which became effective in September 1969. I was fortunate in walking straight into employment with Cathay Pacific Airways in Hong Kong and by the end of the year was a co-pilot on the Convair 880, a four-engine passenger jet similar to the Boeing 707. After 18 months I was made Captain and spent the next eight years progressing up the aircraft levels, through the Boeing 707 to the Lockheed L1011, flying in an area bordered by Korea, Japan, Australia and Bahrain. At Cathay, like other companies I was to fly with where there were Australians on staff, I was subjected to ridicule about my military service. None of the mockers had had combat experience but had climbed to managerial and training positions by the social ladder route. I would often receive challenging comments from these megalomaniacs such as, 'You have a Distinguished Flying Cross, let's see some distinguished fucking flying'. With companies that didn't have any association with Australia, I never had my military service rubbed in my face as something to be ashamed of, and I was allowed to progress without being held back for promotion. In fact, with Saudi Arabian Airlines I became a check airman on the L1011 within four months of joining the company.

Over the years, I have flown for a number of other airlines, including Royal Jordanian Airlines (Jordan), Air Lanka (Sri Lanka), British Air Tours (Manchester), Olympic Airways (Greece) and Caribbean Airlines (Barbados). For three years I flew Kerry Packer's

Lear Jet in what can only be described as a very interesting period of my aviation career. The best company I flew for was Cargolux Airlines based in Luxembourg. It was a no-nonsense airline without personality conflicts and we used to fly all over the planet. Besides the cargo 747s, Cargolux had eight Boeing 747 passenger aircraft operating under the banners of Caribbean Airlines and Lion Air. We would be seconded to Air France, KLM and Lufthansa when flying international passenger and freight routes. On some of the charters out of Manchester and Paris, we could spend up to a week at resorts in the Caribbean and Florida. I don't know how I tolerated it! This was the most demanding, but rewarding and enjoyable flying of my career. I first joined Cargolux on a three-month contract that was renewed repeatedly, eventually resigning after four years of service. My decision to resign was influenced by my family. The job was really very much a single man's life. Living out of a suitcase I would be lucky if I got back to see my family in Australia once every two months. I had two school-age children and when the big pilots' strike of 1989 occurred, there were permanent jobs going in Australia and I had to make the change. So I applied to join Ansett Airlines and was accepted into the worst airline job of my career.

Initially my job with Ansett was good as they were in desperate need of experienced pilots. We flew all over Australia, which suited my long-haul flying background and preference for variety. Promotion to training captain status was quick but soon slowed down as the real 'scabs', those who broke the picket line for personal gain, took over management. I soon found my experience was resented and I started to hear comments such as, 'He thinks he is some sort of war hero—I'll teach the bastard a thing or two!' The tall poppy syndrome was alive and well! It is of significance that many of these managerial pilots had previously failed the pilot training course in the military and not graduated. Perhaps they resented anyone who had achieved what they had not.

In 1993 Ansett was awarded the contract to assist the new Air Vietnam with the structure of its airline and the introduction of the Boeing 767. Having been to all the airports Air Vietnam

THE STRUGGLE TO GET HOME

intended to operate out of, I was selected to familiarise the Ansett Boeing 767 fleet captain, Graham Stewart, with each of them. After doing this, he then cleared me to operate into them. Everything in order! I had about six months attached to Air Vietnam operating out of Ho Chi Minh city. It was a very nostalgic voyage for me and I took the opportunity to visit many of the areas where I had seen combat. The people were very inquisitive about any Europeans and generally very friendly. I found only one ex-VC who would have nothing to do with me. The visit back to the crash site of 18 August incident was extremely emotional. At the point where Colonel Archer and I hid in the water stream there was now a grass one-room hut about five square metres in area. Living in it was a family of six and a pig. The pig was very valuable to the family and they kept it inside. They entertained us with tea and a bowl of rice. I was sitting on the family bed with the pig nuzzling me like a dog. Our conversation was assisted by an interpreter. They were extremely interested in why I was there but I don't think they fully comprehended the death and destruction that had occurred on their back doorstep 25 years earlier.

When Ansett started flying the Boeing 747, anyone who had Boeing 747 experience was held back and discounted while the managerial pilots clamoured to fly the 'big-bird'. At the time, I had had over 5000 hours on the aircraft worldwide, including all of the destinations Ansett intended to fly the 747. The manner in which they were operating the aircraft could only lead to disaster and all the letters I wrote to management on the matter were ignored. I eventually wrote stating that they would have a serious incident if they continued operating the aircraft as they were. Three days later Ansett landed a 747 at Sydney airport with the nose wheel retracted and I really became a target of the megalomaniacs.

Ansett eventually allowed me to fly the 747 but kept me out of the training field. So intent was Ansett management on keeping me from having any influence on the aircraft operations that at one stage they had a first officer, who was about to start his 747 training, fly with me to learn procedures. He was filling in time while he was waiting for his appointed training captain to complete his own check out on the 747.

The 747 were the most versatile and enjoyable airliner I ever flew. I found most of the crews with Ansett a delight to fly with. As soon as I turned 60, the management took me off the 747 under the pretence that I could no longer fly overseas. They put me back on the domestic Boeing 767, which was a sad end to my flying career.

I lasted seven months doing the Brisbane to Sydney and return twice a day run. Although the new management team of the 767 aircraft was a breath of fresh air after my experience with the 747 coterie, I found the repetitive 'bus driver' job soul-destroying. Suffering from PTSD (post-traumatic stress disorder) did not help me tolerate the stress of dealing with management and my pilot licence was medically withdrawn. I could ill-afford to retire but my fate was written in the sands, so to speak. After my sick leave ran out, Ansett laid me off, six months before they went bankrupt. The inevitable decline of Ansett because of poor management was obvious to me for three or four years before they did go under.

I then concentrated on my macadamia/avocado farm but it did not work out. Now I am retired on the menial Government pension, and apart from the odd demon in my head and the continuing battle with the government about their attitude towards US awards to Australians, I am enjoying life pottering about home and flying war-birds at air shows whenever I can. As Lee Alley, of the 5th/60th would say, 'Life is good. The sheets are clean and they ain't shooting at us any more.'

APPENDIX

THE MEDAL OF HONOR ISSUE

WHEN I LEFT THE Air Force in October 1969, I flew passenger jets for Cathay Pacific Airways. After a time I did not give much thought to my experiences in Vietnam. I considered I had recovered from the war and went about my normal life. In 1974 it appeared in the press that General Ewell had recommended me for America's highest award, the Medal of Honor. He wanted to know why it had not been processed. I was surprised, as nothing had ever been said to me about this. It was then that Major Nelson's reference to the 'big one' in August 1968 finally made sense.

In a parcel my mother received from Cathay Pacific, which had been forwarded from the United States Air Force, there was a considerable number of documents and statements, including a Bronze Star citation. As it turned out this interim award was for the action on 18 August and should have been awarded in the field in 1968. It was not because the US recommendations for awards were being unwittingly sent to the Australian Military Command at the Free World Headquarters in Saigon and were never sent back to the originators in the United States. This happened to all foreign awards to Australians who served in the Vietnam War.

There are some 430 recommendations for US awards and literally hundreds of RVN awards for Australian Vietnam Veterans laying dormant in the archives in Canberra. Although the government claims to have changed their policy on foreign awards, they will still not release details of these awards to the recipients and refuse to correct their previous actions.

Prompted by newspaper releases between July 1974 and September the following year, the Australian Government carried out an investigation and claimed that the recommendation for the award of the Medal of Honor on me was never made. The Australian Government at the time was a Labor Government that was opposed to Australia's involvement in Vietnam. On the 25 September 1975 General Ewell and Colonel Patrick advised the Australian Embassy through the US Military that they had indeed made the recommendation. If the Australians were not prepared to act on the United States recommendations then America would. They awarded me the Air Force Cross, which is the highest Air Force award they can present to a foreigner. The actual conferral of the AFC was deferred until the supporting documents could be sent from Australia.

The Australian Government did not send the eyewitness statements. Had the government sent all the eyewitness reports, the US would have processed the award, thereby destroying our government's statement that they had not received the recommendation. I cannot understand why the Australian Government is so afraid of United States military awards. We have fought shoulder to shoulder with the Americans now in several wars, the last of which is in Iraq. Australia owes its freedom from Japanese occupation to the Americans.

While I was a member of the RAAF I did not concern myself about the rights and wrongs of the Vietnam War. The government of the day had assessed the situation and had decided to deploy troops in Vietnam. That was all there was to it as far as I was concerned. I was there to do what my superiors, and the Australian people whom the government represented expected of me. I was disappointed about the way I was treated when I came

APPENDIX—THE MEDAL OF HONOR ISSUE

home after a gruelling eight months in Vietnam and I have since found that many Vietnam Veterans felt the same way.

When I heard that there was no documentation on my service records about the events of 18 August I was surprised. I decided to find out what was on my file using the Freedom of Information legislation. It was not a burning issue for me, just something I wanted to find out about. I know my mother had sent documents on to the defence department in 1975 but I did not know exactly what the documents were as she had no way of producing copies. In July 1985 the US advised that they were unable to locate any documentation on me because details of foreigners had not been placed on computer and they had sent any copies they had to me in 1974. Once the US advised that they could not process the AFC without the missing statements, the Australian Government then dropped the matter altogether. The government had achieved what they had set out to, that is to prevent any American recognition of efforts on the 18/19 August 1968. By this time I had gathered a following of people who had become interested in my case.

In June 1997 General Ewell wrote to Bronwyn Bishop who was then Minister for Defence, Industry, Science and Personnel (see Document A). In it he reiterated his request that I be awarded Australia's highest award, the Victoria Cross of Australia. The Honorable Bronwyn Bishop asked the Director of Honours, Employment and Administrative Policy, Group Captain A.L. Blyth to look at General Ewell's request. Blyth came up with the idea that there should be a Review Panel and that it was important the panel appointed should be seen as objective and independent (see Document C). In a handwritten note to an undisclosed person, but probably Wing Commander McDonald, Blyth went on to say that the chairperson should not be Air Force—why not?—and not be part of Department of Personnel and Equipment (DPE). He suggested a brigadier (E), that is, one-star rank with an engineering background—why an engineering background? He said that the second person should be a senior public servant from the Defence Department, preferably from the Department of Administrative Services, and that the third person should be the person to whom he was writing. Group Captain Blyth went on to ask this person to consider the 'how to' aspects of the review. He also admonished that the review should be done properly otherwise it would only have to be done again.

Wing Commander G.G. MacDonald was the deputy director of Honours and Awards, Department of Defence in Canberra in 1997. An unknown person or persons wrote the terms of reference for the review. It seems that the Department of Defence was reluctant to give its imprimatur to the review out of fear that it would open the door to other similar considerations. The Department of Defence may also have had something to hide and were fearful that the review panel would expose it. In any case the minister approved the appointment of a review panel on 27 October 1997. The panel consisted of Brigadier K.J. O'Brien CSC (Rtd.) and Mr Paul O'Neill, a public servant from the Awards and Symbols Office of the Department of Defence. There was no third member. I gave Barrister K.J. McGhee LLB, himself a Vietnam Veteran, copies of the terms of reference and asked him to look at their suitability. Here is what he said:

Administrative Law which governs the appointment of Tribunals or Review Panels such as that here appointed, requires legislative authority for such appointments. Usually, such bodies are established pursuant to the terms and provisions of some Act of Parliament or subordinate legislation identified as Regulations enacted under a specific Act. This authorising parent Act or and any relevant subordinate Regulation establishes the legislative authority for commissioning a particular Tribunal/Review Panel as well as defining its allotted tasks, so that it can effect the purpose for which it was conceived. This authorising Act and/or its subordinate Regulation usually contains the authority to investigate, as well as the authority to summons witnesses, subpoena documents, receive evidence and generally conduct its review in a particular fashion so as to complete its allotted task according to the terms whereby it has been constituted.

The panel were given the duty to 'ascertain if the US Air Force made a recommendation to the Australian Forces in Vietnam, at any time, that Flt. Lt. G. G. Cooper be recommended for an Imperial gallantry award for the action of 18/19 August 1968'.

The Panel's findings were:

> *The Review Panel was given access to relevant files by the Australian Department of Defence. There is no evidence held on the Department's files to suggest that the US Air Force had made a recommendation to the Australian military authorities either in Vietnam during the conflict or later in Australia. The files make it clear that the issue was first brought to the notice of the Australian military authorities on 21 November 1974 by Flt. Lt. Cooper's mother, Mrs. E. Cooper, in a letter to the then Minister for Defence.*
>
> *This conclusion is supported by the actions of the US military authorities in Vietnam. Following the downgrading of the recommendation for the award of the Medal of Honour (sic) to the Bronze Star Medal for Valor, the matter was treated as finalised until the Australian military authorities drew the matter to the attention of the US military authorities in 1975 as a result of Mrs. Cooper's representations. None of the material made available to the Review Panel in relation to the handling of the issue by the US military authorities suggests otherwise.*

It is untrue that the Australian military brought the matter to the attention of the US military. The converse is true. It was the US military that was seeking information on the events of the 18 and 19 August from the Australians so that they could award me the US Air Force Cross. The recommendation was not downgraded to a Bronze Star as the review panel claimed. As Colonel Patrick pointed out in the recommendation, the Bronze Star was the limit of his authority under the existing circumstances. The Medal of Honor takes several months to process. In the meantime, Colonel Archer awarded an immediate Bronze Star himself as an interim to the processing of the Medal of Honor. This is normal procedure. The award of the Medal of Honor could not be made to me because I was a foreigner; however, because our American commanders in Vietnam were treating the Australians as equals, the Medal of Honor recommendation was made.

General Ewell today has said that if he knew what was going to happen, he would have handled the matter quite differently in 1968. The Americans do not make the award of the Medal of Honor without a thorough investigation into the incident for which it is to be awarded. The Americans sent me the citation to the Bronze Star in a package along with other documentation, which included seven eyewitness accounts of the events that took place on the 18 August 1968. My mother forwarded the documents to Lance Barnard, then the minister for Defence on 3 July in 1974. My mother was of the 'old school' who trusted politicians and believed they were there to help.

Under the Freedom of Information legislation I viewed the Department of Defence files on myself. The folio numbers are not sequential, indicating that documents are missing. The files show that the recommendation in question was on file before my mother sent the documents she had received from the USAF via my employer in Hong Kong. The documents sent by my mother are annotated 'received from Mrs. E. Cooper'. There are copies of the same documents that are not annotated, indicating that they were on file already. If the panel was given access to the same files as myself then they were viewing already censored files. Not only that, but any panel with a little gumption could have drawn the same conclusions as I have drawn. That they did not doesn't mean they had a mediocre IQ or they chose not to do so but the bottom line is that the findings of the panel are blatantly wrong. The minatory side is that they may have set out to deliberately come to that conclusion. In fact they may have been instructed to do so.

The whole of the panel's deliberations, its conclusions and its arguments become unravelled at this point. The only thing that the panel should have considered is whether or not I had been recommended for the Medal of Honor and whether the Americans had asked the Australians to award me the equivalent Australian honour. Despite Group Captain Blyth's admonishment to Wing Commander MacDonald to get it right, he did not. In fact he could not have got it more wrong. Group Captain Blyth told

APPENDIX—THE MEDAL OF HONOR ISSUE

MacDonald that he should make up the third person on the panel. For some reason he did not. Brigadier O'Brien was a retired artillery officer who had no combat experience of the type in which I was involved on a daily basis. Mr. Paul O'Neill, as far as I am aware, had no combat experience at all.

In Clause 2(a) of the Terms of Reference the panel was to 'ascertain if the action of Flt. Lt. G. G. Cooper, on 18/19 August 1968, took place as described' (see Document D). This clause is irrelevant and should not have been included unless they distrusted the word of several senior US officers. The Americans had authenticated the action 30 years earlier. General Ewell wrote the recommendation for the Medal of Honor. General Ewell is a military commander who has proven himself in many areas of the US Military for over 40 years.

The panel's findings cannot be published in full here, but page 3 of the report states:

> *The Review Panel approached its task by seeking to identify or discover material available at the present time which can substitute for corroborative statements that would, in the normal course, have been made at the time the recommendations were made. The Review Panel has not set out to discredit or discount any of the material made available to it, but to make its own assessment of the material on the simple test of whether it is conclusive and enables all of the elements of the incident of 18 and 19 August 1968 as described and all circumstances surrounding the incident to be free from any doubt.*
>
> 1) *The incident of 18 and 19 August 1968*
> *The incident, involving the crash of the helicopter on 18 August 1968, death of the pilot on 18 August 1968 and helicopter rescue of Flt. Lt. Cooper and a US Army Officer on 19 August 1968, is described in:*
>
>> *a) a narrative attached to a recommendation by Maj. Gen. Julian J. Ewell, Commanding Officer, US Army 9th Infantry Division dated 20 August 1968 that Flt. Lt. G.G. Cooper be awarded the Medal of Honor for his gallantry in action on 18 and 19 August 1968 (Appendix 1). Recent support for the recommendation*

and narrative is given by General Ewell on 20 May 1997 and 13 February 1998 (Appendix 2).
b) a recommendation prepared by Lt. Col. James T. Patrick, Commanding Officer 19th Tactical Air Support Squadron, 7th Air Force, USAF dated 23 August 1968 that Flt. Lt. G.G. Cooper be awarded the Bronze Star Medal for Valor in lieu of the Medal of Honor as the latter award could not be awarded to Flt. Lt. Cooper, a foreign national (Appendix 3). Recent support for the statement recommending an award is given by Colonel William L. Walker USAF Rtd. who prepared the recommendation for signature by Patrick (Appendix 4).
c) a statement by Colonel James Hoag dated 19 August 1968 relaying an account of two F4 pilots who had seen the helicopter crash from their aircraft flying overhead (Appendix 5).
d) a statement prepared by Flt. Lt. Cooper on 12 January 1999 (Appendix 6).
e) a diary entry prepared by Flt. Lt. Cooper (Appendix 7).
f) a citation to accompany the award of the Bronze Star Medal for Valor in respect of the 18 August 1968 incident (Appendix 8).

In paragraph (b) the Bronze Star was not awarded in lieu of the Medal of Honor but as an interim while the Medal of Honor was being processed (see Document F). In paragraph (c) there were four eyewitnesses since both F4 Phantoms also flew with a back seat radar operator.

The panel said that my mother provided the citation to the Bronze Star for Valor. This being the case, then the Defence Department must also have received all the documents my mother forwarded to Lance Barnard in 1974, which included seven eyewitness accounts that had been prepared by the US military to support their recommendation that I be the recipient of Australia's highest award (see Document G). I have never received the Bronze Star for the incident on 18 August 1968. In fact I had never heard about it until the panel came up with the citation. The panel is here setting out to authenticate the action of the 18 and 19 of August 1968 30 years after the event. The panel's report continues with:

APPENDIX—THE MEDAL OF HONOR ISSUE

b). The incident as described
  i. Recommendation by Major General Julian J Ewell (Appendix 1)
On 20 August 1968, General Ewell recommended the award of the Medal of Honour [Honor] to Flt. Lt. Cooper for his gallant actions on 18 August 1968. The letter of recommendation is accompanied by a narrative prepared by Colonel William L. Walker. The recommendation and narrative were provided to the Department of Defence, Canberra, on 21 November 1974 by Mrs. E Cooper. The relevant components of the incident as described in General Ewell's recommendation and narrative are:
- the pilot of the command helicopter was shot dead and Flt. Lt. Cooper controlled the descent of the helicopter to reduce its impact with the ground;
- a Brigade Commander, a passenger in the helicopter, was hit in the neck with a bullet which also struck Flt. Lt. Cooper's flying helmet;
- the crash occurred in the late afternoon;
- the crash occurred at a point of contact between friendly and enemy forces, some 200 metres in front of the enemy line;
- under fire, Flt. Lt. Cooper assisted the Brigade Commander to safety;
- Flt. Lt. Cooper repelled enemy troops and killed at least 10 at close range;
- the Brigade Commander and Flt. Lt. Cooper exposed themselves to enemy fire to escape;
- Flt. Lt. Cooper exhausted his ammunition while covering the Brigade Commander who was being hoisted aboard the helicopter;
- with a now empty handgun, Flt. Lt. Cooper defended himself against attack by two enemy troops who he killed; and
- Flt. Lt. Cooper saved one of the 9th Division's most valuable Brigade Commanders who was wounded.

ii. Recommendation by Lt. Col. James T Patrick (Appendix 3)
On 25 August 1968, Colonel Patrick recommended the award of the Bronze Star Medal for Valor in lieu of the Medal of Honor as the Medal could only be awarded to US citizens. The factual components listed in i) above are repeated. The recommendation and narrative were provided to the Department of Defence, Canberra, on 21 November 1974 by Mrs. E Cooper [see Document H].

iii. Statement by Colonel James H. Hoag (Appendix 5)
On 19 August 1968, Colonel Hoag relayed an account of two F4 pilots who had seen the helicopter crash from their aircraft flying overhead. The statement was provided to the Department of Defence, Canberra, on 24 February 1997 by Mr. F Kirkland who had obtained it from Mr. G.G. Cooper on 19 December 1996. [see Document I]

The relevant components of Colonel Hoag's statement are:
- The helicopter containing Flt. Lt. Cooper crashed on 18 August 1968 at Cai Be;
- Flt. Lt. Cooper, distinctive by his Australian flight suit, was seen half carrying an Infantryman under hazardous conditions.

Not considered is the fact that I was talking with the F4s immediately before being shot down so they would have been well aware that they were talking with an Australian.

As a result of the Review Panel's inquiries, the HQ USAF advised during 1978 that Col. Hogan [Hoag] had been approached previously on this matter some two years earlier but had no recollection of the incident. Col. Hoag had indicated that he could not materially add to the investigation.

iv. Statement by Flt. Lt. G.G. Cooper (Appendix 6)
On 12 January 1999, Flt. Lt. Cooper provided a written statement to the Review Panel recalling his account of events on 18 and 19 August 1968. Flt. Lt. Cooper states:
- the helicopter, an OH23 Raven Scout, crashed late in the afternoon of 18 August 1968;
- he did not take control of the helicopter during its grounding, rather he switched of (sic) the ignition after the helicopter hit the ground;
- confirms the death of the pilot;
- he extracted the Colonel from the helicopter and removed him away from the helicopter;
- confirms that he killed two enemy but unable to confirm any others;
- confirms that he struck two enemy soldiers from behind with his empty CAR15 carbine;

APPENDIX—THE MEDAL OF HONOR ISSUE

– the incident occurred at the 1st Brigade contact area, some 6km south of Can Giuoc; and
– he directed 2 F-4 aircraft over the contact area and recalls the F-4 aircraft passing overhead after the helicopter crash.

v. *Flt. Lt. Cooper's Diary Entry (Appendix 7)*
In response to a request from the Department of Defence, Canberra, Flt. Lt. Cooper on 15 June 1999 provided the Review Panel with a photocopy of pages of his diary for 18 and 19 August 1968. The crash and evacuation described in iv) above are supported.

vi. *Bronze Star Medal Citation (Appendix 8)*
On 3 July 1974, Mrs. E. Cooper forwarded to the then Minister for Defence a copy of the citation to the Bronze Star Medal awarded to Flt. Lt. Cooper in respect of the 18 and 19 August 1968 incident. The factual components listed in i) above are summarised with minor variations as indicted below:
– the crash occurred on dusk on 18 August 1968;
– Cooper fought off several enemy attacks throughout the night; and
– Cooper killed a further two enemy as they tried to overpower him.'

The Review Panel continued to refer to the documents provided by my mother. All of these documents were on file prior to my mother sending the copies she received from the US military. It is almost a suggestion that she created them. On 5 March 1974, the US retired all documentation they held on foreigners who served with them during the Vietnam War. My mother simply forwarded them on to Lance Barnard in 1974 after having received them from my employer, Cathay Pacific Airlines (the Americans had sent them to me there). I was on leave in Europe at the time so Cathay posted them to my home address in Australia.

In the summary of General Ewell's narrative of the incident there are some minor differences to my account of the incident in chapter 9. The reason for this is that General Ewell had to write an account of the event without reference to me. He would have asked Colonel Archer when he returned to Dong Tam on 19 August 1968. Colonel Archer was unconscious before the Raven crashed. The only thing he remembered was flying at

200 feet (60 metres) prior to crashing. Colonel Archer thought that I had taken control of the helicopter and landed it safely. This was not the case as you read in chapter 9. The fire that brought us down was directed from the main contact area. If you look at the map you can see that we crashed some 500 metres west of where we were hit. The pilot was still at the controls and was either dead or unconscious. One of his feet must have pushed down on the directional control because we were spinning.

Later when the report says that we exposed ourselves to enemy fire to escape, Archer was still incapacitated. I dragged him clear of the Raven and through long grass to the canal bank. We did not have to expose ourselves because we were already being fired upon. The enemy could not draw a clear bead on us, as we were hidden by the long grass after I dragged the colonel clear of the aircraft. Had the grass in the rice paddy been a little shorter we both would have been shot.

In dealing with the recommendation by Colonel Patrick the panel again refers to the fact that the documents were provided by Mrs. E. Cooper. The documents were already on file as I have said. The documents had been on the Department of Defence file since August 1968. In Colonel Hoag's statement he refers to a relayed account of the incident by the two F4 Phantom pilots who were over the scene. If you remember from chapter 9, I was in the process of directing them when we were shot down. The statements of the F4 pilots were one of the seven missing eye-witness accounts of the incident that the US military would have forwarded to the HQAVF in 1968. The statements had been used in support of General Ewell's recommendation that I be awarded the Medal of Honor. It is not surprising that Colonel Hoag could not remember. He was the operations officer and he signed dozens of letters every day. And 30 years had passed.

The report continued:

c). *The elements of the incident.*
In the absence of first hand accounts or eyewitness statements prepared at the time of the incident, other than that of Flt. Lt. Cooper, it was necessary

APPENDIX—THE MEDAL OF HONOR ISSUE

for the Review Panel to seek to establish whether support for the incident could be obtained from other sources. The absence of supporting material prepared at the time of the incident, including statements by eyewitnesses or other persons or events which corroborate Flt. Lt. Cooper's first hand account, requires the discovery of factual material which is directly relevant to the incident or which leave no doubt as to the circumstances surrounding the incident.

Here they refer to the absence of first-hand accounts. Some of the eyewitness accounts were made absent by the Defence Department. General Ewell would not have made his recommendation without making sure of the facts. Others would have collected eyewitness accounts of the event. In fact there were seven of them. These seven statements were part of the recommendation received by 504 TASG and filed with HQAFV in Saigon sometime after 19 August 1968. The Americans made copies of the seven statements. Once they retired the documents on foreigners that they were keeping, they sent mine to me at Cathay Pacific. On my file there should be two sets of the eyewitness statements. You can see that these seven statements were with the documents sent to Lance Barnard in 1974. When I examined my file under FOI the statements were not there. A person or persons unknown, with or without the imprimatur of the Department of Defence, had removed documents from my file.

The report continued:

The Review Panel has identified a number of elements of the incident and has sought to determine whether they are conclusive and would corroborate the incident as described.
i. The command helicopter was shot down on 18 August 1968.
A first hand account of the helicopter crash is provided by Flt. Lt. Cooper in a statement dated 12 January 1999 (see Appendix 6). It is stated that the incident occurred late in the afternoon of 18 August 1968, some 6 kilometres south of Can Giuoc and in the vicinity of the Ist Brigade contact area when the command helicopter was shot down from low altitude by enemy gunfire. The helicopter had been operating during the day in

269

support of the 1st, 2nd and 3rd Brigades of the 9th Infantry Division, US Army. The helicopter is identified as a Hiller OH23 Raven Scout which may have been attached to a unit outside the 9th Infantry Division.

At the time of the incident, the Ist Brigade conducted operations with the following units:

2nd Battalion, 39th Infantry
2nd Battalion, 60th Infantry
5th Battalion, 60th Infantry (Mechanised)
1st Air Cavalry Troop (A/3-17)

The 1st Brigade Combat Operations After Action Report for 18 August 1968, reports that the 2/60th Battalion first made contact in southwestern Gan Giuoc District at XS8264 on 18 August 1968. It further reports that elements of the 2/39th were also inserted to encircle the engaged enemy unit.

Daily Staff Journals/Duty Officer's Logs for 18 August 1968, reveal references to a number of incidents involving enemy action against helicopters during the afternoon of 18 August 1968:

- '1429 (CBS 2-60 rpt 1 slick hit by RPG; down at this time; fire from XS828648' (Daily Staff Journal, 18 August 1968, 1st Brigade, 9th Infantry Division, Item No. 26 and 29);
- '1520 (C) 2-60 rpt 2 WIA as result of chopper being hit; called dust off.' (Daily Staff Journal, 18 August 1968, Ist Brigade, 9th Infantry Division, Item No. 32);
- '1638 (C) 2-60 rpt I KIA, 2 WIA result of chopper incident (2 WIA reported earlier)' (Daily Staff JournaL 18 August 1968, 1st Brigade, 9th Infantry Division, Item No. 36);
- '1735 (C) 2-60 rpt A/3-17 chopper shot down 1000m NW of contact area by ground fire vic 815665' (Daily Staff Journal 18 August 1968, lst Brigade, 9th Infantry Division, Item No. 40).

In the 1st Brigade Combat Operations After Action Report for 18 August 1968 a reference is made to what appears to be the helicopter at the 1429 reference above:

- 'Z at XS 828646 was hot and 1 helicopter was hit by an RPG and caught fire. Result was 2 US WIA which were dusted off.'

Later in the same After Action Report, the helicopter at the 1735 reference above was described, as was its subsequent recovery on 18 August 1968:

APPENDIX—THE MEDAL OF HONOR ISSUE

- 'At 181735H 1 Cobra of A/3-17 was hit by SA fire and forced to land at XS815665. In addition 1 Cobra and 1 LOH were hit by SA and flew to Tan An to estimate damage. Cobra downed at XS815665 was evacuated by CH47 at 181845H'

The After Action Report did not however make any reference to the helicopter at the 1520 and 1638 references above. It is possible that they refer to the helicopter in the 1429 report rather than a separate incident.

An examination was made of the Operational Notes for August 1968 for the 191st Assault Helicopter Company which provided support to the 3-60th Battalion of the 2nd Brigade, 9th Infantry Division on 18 August 1968. A Command and Control helicopter of the 191st Company was forced to land at 1700 on 18 August 1968 as the result of enemy fire. The helicopter was described as unflyable with no injuries or loss of personnel reported. Additionally, it is clear from the grid coordinates of the action that it occurred far distant from the incident reported by Flt. Lt. Cooper.

General (sic) Emerson (Commanding Officer, 1st Brigade, 9th Division until 26 August 1968) 'remembers what he refers to as a 'scout helicopter' having been shot down on 18 August 1968 in the same area as the one in which Flt. Lt. Cooper was a passenger, he can recall nothing further'.

We were under the Command of Colonel Emerson for this mission. There were six helicopters brought down around the contact area within 20 minutes. Little wonder there may have been some confusion with so many helicopters hit at once. There is reference to an LOH and a Cobra being hit by small arms fire and flying to Tan An for assessment. The panel did not follow this up. The LOH helicopter referred to above was almost certainly us. The report above does not say that the LOH returned to Tan An. If you look at the map you can see that we spiralled away to the west and across the Rach Cac River before crashing. As we were there one second and gone the next, it may have been assumed that we departed for Tan An.

The panel continued to sift through documents and to try to gather information for 12 months trying to corroborate the facts of events that occurred 30 years earlier. Here is what they said about the statements of the F4 Phantom pilots:

A statement on behalf of F4 pilots who claim to have seen the crash and Cooper moving across the ground after the crash is relevant. However, given the operating height of the aircraft, the Review Panel is unable to accept this evidence as conclusive, thereby overriding material which contradicts the report or providing substantive proof to complete the record which is otherwise silent or inconclusive. It is not a first hand or eye-witness report.

This statement displays the panel's ignorance of operations of this kind. The role of the FAC is to do precisely what the panel claims is not possible. Neither of the panel's members had ever been involved in this kind of operation. Any efficient pilot is capable of distinguishing personnel on the ground. I can say this with authority. Had we not been skilled in identifying personnel from the air we would have bombed our own troops. The Phantom pilots were experienced and flew sorties of this type on a daily basis. They were not flying over at 30 000 feet (9000 metres) but were down low and slow. Not only that, both pilots were talking with me just prior to us being shot down. In fact, I was in mid-sentence when hit. They would have recognised my Australian accent. It is standard practice for the F4 crews to be debriefed after the mission by their operations officer to correlate and centralise data for intelligence. The operations officer's statement then is the equivalent of the four eyewitness accounts of the events on the ground.

The Review Panel is also satisfied that the helicopter grounded at 1429 can also be excluded on the basis that it was forced to land early in the afternoon, rather than late in the afternoon as described. In addition, the two wounded personnel were evacuated soon after the grounding.

In respect of the 1520 and 1638 references to the helicopter being hit by enemy fire, the Review Panel is unable to form a conclusive view as to whether it is or is not the helicopter described in the incident involving Flt. Lt. Cooper. Uncertainty is caused by:
  – the report refers to the helicopter being hit, rather than grounded;
  – the possibility that the helicopter reported as being grounded at 1429 is the subject of the reports;

- *whilst other reports from other contemporary sources refer to the 1429 and the 1735 groundings, no mention is made of the 1520 and 1638 hits, nor of a consequential grounding of a helicopter on 18 August 1968; and*
- *the times do no[not] correspond with a late afternoon or dusk incident as described.*

The Review Panel is unable to be satisfied that the material made available to it conclusively supports this element of the incident.'

The Panel set out to verify whether the action took place. But the Americans had already verified the action and an interim award had been made. The brigade commander I dragged from the downed helicopter was Colonel Archer. He was the 2nd Brigade Commander but he unfortunately died on 23 November 1978. The material in the panel's report drags on for some 25 000 words.

ii. the pilot of the helicopter was killed:
The identity of the pilot cannot be established.
In submissions put to the Review Panel it has been established that no record of the death of the pilot can be found beyond that recorded in Flt. Lt. Cooper's diary.

The death of the pilot is recorded in the recommendation written by General Ewell.

The Review Panel asked therefore that US military records be examined by Australian Defence Staff attached to the Australian Embassy, Washington. US military records state that only one US Army Warrant Officer was killed in Vietnam during the period 17 to 19 August 1968. W3160336 James Doyle Eisenhour from Lacrosse KA, born 10 October 1946, married, was killed on 18 August 1968. WO Eisenhour was the pilot of a UHIH helicopter shot down in Vietnam Military District Three (Hau Nghai), in the vicinity of Hiep Hou. The passengers on the helicopter were listed as Maj. Farmer and CW2 Chandler. WO Eisenhour was assigned to the 240th Assault Helicopter Company at the time.

In the absence of the pilot of the helicopter involved in the Cooper incident being identified, the Review Panel is unable to be satisfied that this element of the incident is conclusively supported by the material made available to it.

I stated that the pilot with me was killed and the panel has accepted my statement here as it suits their argument. Throughout the whole review the panel accepted my statements only where it supported their required findings. I said that a helicopter hovered over the wreckage about 30 minutes after we were shot down. We were too far away to see if the crew of this helicopter extracted the pilot. Although the pilot's wounds were horrendous, I know of examples where similar cases have survived for days and recovery dependent upon how quickly medical attention was supplied. The helicopter pilot may not have been an American. He could have perhaps been Canadian. In this case he would not be listed on the American KIA list. Further, as he was from a unit outside the 9th Infantry Division, his death would not be listed necessarily on 18 August 1968 as he would not be listed as missing immediately. If he was listed as MIA, it could be weeks before he is noted as KIA. The date of death is often different to the date when wounds are inflicted. For example W01 Michael Ray Harlamert was recorded as KIA on 20 August 1968 but his death was from wounds received on 18 August 1968 in Hua Nghia province. The records show the date of confirmed death, not the date of the incident.

Once again, the Panel has displayed a lack of appreciation of the situation.

d). Other Matters
A number of matters were raised with the Review Panel which it has addressed prior to making its findings.
i. The Location of the Crash Site
The contact point of the 1st Brigade on 18 August 1968 is stated in the 9th Division records created at the time as occurring at map coordinates XS8264, being approximately 2-2.5 km south of Ap Tay Phu, east of

APPENDIX—THE MEDAL OF HONOR ISSUE

the Song Rach Cac River, within Long An Province. Flt. Lt. Cooper states in his diary entry for 18 August 1968 and in his subsequent statement, that the crash occurred some 6km south of Can Giuoc which places the crash site within the vicinity of the 1st Brigade contact point as stated in the 9th Division records.

Further, in a map described as being drawn by Flt. Lt. Cooper in 1979, the crash site is shown as occurring east of Highway 5, south east of Rach Kien and approximately 20 km south of Saigon. The location is consistent with the general vicinity of the 1st Brigade contact point

The Review Panel noted in a number of submissions that the location of the crash site was inconclusive. The Review Panel appreciated that those making the submissions may not have wished to exclude the possibility of the crash occurring in support of other elements of the 9th Division either elsewhere in Long An Province or in another province. The issue is important in attempting to identify the 9th Division Brigade Commander in the event of him operating in support of his Brigade at the time of the crash.

The location of the crash site is also confused by the statement by Col. James Hoag on behalf of the F4 pilots who claim to have seen the crash (Appendix 5). It is claimed that the crash occurred at Cai Be near Rach Kien. Both locations are far distant from each other. However, Flt. Lt. Cooper in his diary, places a village named Cai Be within the vicinity of the 1st Brigade contact point. This apparent inconsistency was explained in a submission to the Review Panel that a number of villages were locally named as Cai Be.

The primary documents point to the site as being located within 1-1.5km of the 1st Division (sic) contact point. The relevant 9th Division records created at the time, particularly of those elements of the 1st Division at the contact point, do not corroborate the crash nor do they identify the crash site.

The alternative view put to the Review Panel is that the crash occurred at another 9th Division contact point, for example the 2nd Brigade thereby leaving the way open for Col. Robert Archer to be identified as the unnamed Brigade Commander, either in Dinh Tuong Province or in Kien Hoa Province.

In either case, based on all of the material made available to the Review

Panel, it is not possible to establish with certainty the location of the crash site.

It has not been emphasised that I drew the map the panel was looking at from memory in 1975 before seeing any logs and long before the panel was convened. It has already been established that we were operating in support of all three Brigades of the 9th Infantry Division that day. That is, the 1st, 2nd and 3rd Brigades. The three Brigades were operating over a wide area in the west of III Corps Tactical Zone.

At the bottom of the map, just below grid-square 8265, you can see the 1st Brigade's forward operations area. That is where we landed and I spoke directly with Colonel Emerson not long before we were shot down. He desperately wanted TACAIR to relieve his troops in the main contact area in grid-square 8266. Helicopters were being shot out of the air like flies. The whole plain was a rice-growing area. The area was as flat as a pancake and not the type of country to operate helicopters in relative safety. Helicopters are vulnerable aircraft. We were flying east of the Rach Cac River when the pilot was hit. We must have spiralled down to land about 700 metres from where that unlucky shot hit the pilot.

The Vietnamese repeat placenames. For example there are many places in Vietnam called Tan An. There is also more than one place called Cai Be. I used the name Cai Be near Rach Kien to refer to where we were while I was talking with the Phantom pilots. The pilots had a good view of the whole area. Visibility was still good as it was a clear day. The Cai Be name I used actually identifies which Cai Be is being referred to and clarifies the location. When there are places of the same name they are always isolated by saying which place they are near. For example Ap Bac near Cai Lay as distinct from Ap Bac near Can Giuoc. The Cai Be where the 2nd Brigade were in contact was Cai Be near Vinh Long and is not a village but a sizable town. The hamlet near where we came down was not named on the map so I used Cai Be near Rach Kien in my briefing to the fighter pilots. This was

APPENDIX—THE MEDAL OF HONOR ISSUE

better than referring to it as 'the hamlet on the map without a name'—that would have been tedious and time consuming in an operation where every second is vital.

The panel is erroneously talking about the 1st Division. I was flying in support of the 9th Division, not the 1st Division. The 1st Division operated in III Corps and was the first US Army infantry to arrive in Vietnam. At the time of this incident, the 1st Division was based at Lai Khe. This was at the 1st Brigade (9th Infantry Division) contact point and not the 1st Division, as referred in their report. We were at least one kilometre away from the contact area when we came down. It is not surprising that we were not observed to crash, and so there is no explicit reference in the 1st Brigade logs to a LOH crashing. But there is reference to a LOH being hit and departing for Tan An. Not only that, the soldiers had their heads down and would have been too busy to notice us. The CO of the 1st Brigade, Colonel Emerson, referred to a scout helicopter being shot down but that does not appear in the logs either.

The duty roster shows that I was to fly with Colonel Archer and we were working with boats operated by the Mobile Riverine Force in the Rach Cac River alongside of the 1st Brigade contact area. The boats were forming a blocking operation to prevent the enemy from escaping across the river. The only confusion or uncertainty here is in the minds of the panel. The panel has gone to great lengths to cloud the issue in order to support their negative requirements.

The panel next examined the primary documents; that is the documents containing General Ewell's and Colonel Patrick's recommendations. The person who commissioned the analysis was Mr. Brian Bates, an ex-police officer who had been discharged from the NSW Police Force on medical grounds. Brian Bates had approached me some time previously with an offer of help to write my story.

I accepted the offer because it seemed like a good idea at the time. I was still flying international routes and I was often away from home and had little time to write. As it turned out Brian did not have the stamina to complete the job. At one point he made unwelcome suggestions to my wife when he visited our home, ostensibly to obtain documents for the proposed book. The book was never written and he and I parted on bad terms. This is what the panel said in their report.

iii. The Primary Documents

On 9 December 1998, the Review Panel was provided with a report prepared by Forensic Document Services P/L who had subjected the primary documents being examined by the Review Panel to forensic study. The forensic study had been commissioned by a member of a group of individuals supporting the case for Flt. Lt. Cooper's award to be upgraded, following their identification of what appeared to be exact similarities between the signatures on the primary documents with documents relating to previous awards made to Flt. Lt. Cooper.

In its report, Forensic Document Services concluded that the primary documents describing the incident of 18 and 19 August 1968 contained copied signatures of the nominators. These documents, being the subject of the Review Panel's inquiries, are shown at Appendices 1, 3 and 5. Following receipt of that report, this group of individuals withdrew their support for the case and informed both the Review Panel and Mr. Cooper. It should be noted that Forensic Document Services has also stated that, because the copies of the original documents had been subject to multiple photocopying processes, its findings were not conclusive.

Copies of the above documents and other documents were also subjected to forensic study by John Heath Document Consultations, at the request of Mr. Cooper, and the findings were provided to the Review Panel on 29 March 1998. John Heath Document Consultancies also concluded that, whilst the signatures on a number of documents were likely to be copies of signatures appearing on other documents, its findings were not conclusive.

Having considered the results of the forensic examinations into the signatures, the Review Panel makes no finding as to the veracity or otherwise of the documents as no original documents were examined in either study and consequently the results were inconclusive.

APPENDIX—THE MEDAL OF HONOR ISSUE

Firstly, it was not a group of supporters who withdrew their support. It was one person who used this issue in an attempt to scuttle the enquiry for personal reasons. Brian Bates used the signature matter to leave the support group. He could have left in a dignified and professional manner, perhaps retaining some credibility. It is on record that rubber signature stamps were commonly used because the officers were often in the field. With US forces, the commanders are required to be present in combat as well as to conduct their administrative duties. They keep their rubber signature stamps locked up. As time permits they read through the outward correspondence, typed and collated by their administrative staff. If they approve of the material they stamp their signatures on the documents. This procedure saves a lot of time when there are hundreds of documents to sign daily.

The Forensic Document Service referred to 'cut and paste' procedures. As personal computers did not exist in 1968, 'cut and paste' had not yet been invented and the normal method of copying was by using the Gestetner process. In any case, the Forensic Document Service finding was that they could not come to a conclusion so this part of the Panel's investigation was included only to throw doubt on the legitimacy of the primary documents. The creators of the documents are living senior officers who have said time and again that they raised the recommendation.

I requested John Heath to examine the documents and to make a report in order to counter the intended damage to my integrity. With the exception of Brian Bates my supporters remained steadfast in their belief that the Defence Department had erred in their handling of General Ewell's recommendation. It is interesting that the person who tried to scuttle the investigation for his own private reasons was refunded the costs of the forensic service even though he instigated the analysis of his own volition. I was not reimbursed, despite having to hire the forensic services of John Heath to protect my integrity. It seems that this section was included to support the negative results required by the Panel. The Americans must have wondered what the Panel hoped to achieve. It would have seemed to them that the conclusions of the panel were the results of a straw man argument. In other words the panel knew the conclusions before it started its deliberations. The report continued:

iv. Verification of the incident by the US military authorities
In a submission to the Review Panel, it was revealed that the recommendation made by General Ewell for the award to Flt. Lt. Cooper of the Medal of Honour, the United States highest gallantry award, may have been prepared on a pre-signed document or on a document with a copied signature.

A similar suggestion was made about the recommendation by Colonel Patrick for the downgraded award to Flt. Lt. Cooper of the Bronze Star Medal for Valor, namely that the recommendation was made on a pre-signed and copied form.

In additional material provided to the Review Panel it is suggested that the practice of using pre-signed and copied documents occurred as a matter of course to reduce the administrative burden on military officers required to fulfill their heavy operational obligations.

The Review Panel considers that, regardless of the practice adopted by the US military authorities in Vietnam to initiate recommendations for awards, the practice needs to be viewed in the broader context of the procedures used to process gallantry awards.

The US Seventh Air Force was authorized to approve awards in Vietnam up to and including the Silver Star. Awards above this level—the Distinguished Service Medal, Air Force Cross and the Medal of Honor—were independently processed by the US Defence Department's Awards and Decorations Branch. It was mandatory that all recommendations for awards of the Silver Star and above for foreign nationals be referred to the Branch. All recommendations referred to the Branch were assessed and verified.

Whether a recommendation is prepared on a pre-signed and copied form is of little significance if the independent assessment and verification is undertaken.

For the incident of 18 and 19 August 1968, this independent assessment and verification did not take place at the time as the decision was taken by Colonel Patrick to downgrade the award to the Bronze Star Medal for Valor. The Bronze Star Medal was subsequently awarded to Flt.

APPENDIX—THE MEDAL OF HONOR ISSUE

Lt. Cooper on authority delegated to the US Seventh Air Force in Vietnam based on a recommendation prepared on what is claimed to be a pre-signed and copied form. There was therefore no requirement for the incident to be independently assessed and verified.

The Review Panel is conscious of the fact that the award of the Bronze Star Medal for Valor does not require the same level of assessment and verification as is required for the award of the Medal of Honor. It has also been pointed out to the Review Panel that it was routine for awards of the Bronze Star Medal for Valor to be made as the result of pre-signed and copied recommendation forms and evidence of one such award was provided to the Review Panel.

The US military authorities did however attempt to subsequently verify the incident of 18 and 19 August 1968 between 1975 and 1978. The recommendation, presumably that made by Colonel Patrick, was reprocessed and the incident was assessed against the criteria for the Air Force Cross. The conditions relating to the award state that it is awarded for extraordinary acts of heroism in action against an enemy of the United States, not justifying the award of the Medal of Honor.

The US Department of Defence advised that the award could not be made because of a lack of eyewitness support or the name of the Brigade Commander. (Appendix 9). It was the view of the US Department of Defence that the original recommendations made by General Ewell and Colonel Patrick required independent verification before the award of the Air Force Cross could be made.

As the incident occurred within the operational jurisdiction of the US Armed Forces with the recommendation for the gallantry award to be made to Flt. Lt. Cooper being made by senior US military officers, the Review Panel is required to give due weight to the findings of the US military authorities arising from its reassessment of the original recommendation.

General Ewell and Colonel Patrick have each confirmed their respective recommendations. The independent assessment and verification could not take place, as the Australian authorities in Vietnam never returned the recommendation to the American system so that it could be verified. The Australian authorities subsequently 'mislaid' most of the vital eyewitness reports. The RAAF was collecting all US recommendations on Australians and filing them during the Vietnam War in support of their policy to deny Australians US gallantry awards—not only mine but many others. An independent assessment and verification cannot be confirmed as the Australian Government had all the US documentation and I doubt, after what has happened, that the missing eyewitness accounts will be found.

This is scandalous because there were two copies of these on file; the originals sent by the US to 504 TASG in 1968, and the copies unwittingly sent to Lance Barnard by my mother. Also, if the Bronze Star was awarded then that verifies that the action happened as stated. The Bronze Star was not a downgrade but an immediate award as an interim to a higher award. This assessment and verification does not affect the matter in any way. The panel presumes that the US recommendation for the Air Force Cross was made by Colonel Patrick. Documents state that it was the US Air Force Awards Branch that upgraded the interim Bronze Star (see Document J).

When the US Department of Defense set about verifying the action for the Air Force Cross, they had to ask the Australian Defence Department to send them the documentation as the US had already sent the documents to the RAAF in 1968. The Australian Defence Department only forwarded the basic recommendation and not the eyewitness statements. The reason for this becomes clear when we view the reasons why the US could not verify the action in their consideration of the Air Force Cross. Under the FOI files it can be shown that seven statements are missing from my personal files.

This is a round robin scenario. The Australians created the impasse in the first place and that is why the Americans came up with their decision that the Australians are now using to support their conclusions. The Australian interference with the US documents is a crime, I believe. To suggest that General Ewell, supported by a dedicated awards branch, submitted a recommendation for his nation's highest award in an unprofessional and incomplete format is ridiculous. Finally, here are the Panel's findings:

APPENDIX—THE MEDAL OF HONOR ISSUE

e. Findings

In order to establish that the incident of 18 and 19 August 1968 took place as described, the Review Panel is required to take into account the statements of Flt. Lt. G.G. Cooper and material made available to it from a number of sources including that obtained from the US military authorities. The overall effect of this additional material, when taken in conjunction with the original gallantry award recommendations, is to create a level of uncertainty about the material facts of the incident.

US official records obtained by the Review Panel are either inconclusive or incomplete. Records, which the Review Panel was unable to discover, relate to:

the [identity of] rescued US Army Officer;

hospitalisation of Flt. Lt. Cooper and the rescued US Army Officer; The Panel did not view or discover [any] hospital records.

– the pilot killed, recovery of the body and record of death;

– [the] eyewitnesses; and

– [the] grounding and recovery of the helicopter.

As a consequence, further uncertainty is created about the material facts of the incident.

Of critical importance are the following:
– there is no independent record of a 94 Hiller OH23 Raven helicopter being shot down on 18 August 1968 and of it being either destroyed or recovered;
– the unit assignment of the helicopter cannot be established;
– there is no independent record of a 9th Division Brigade Commander or other Brigade Commander involved in the contact between elements of the 9th Infantry Division and the enemy on 18 August 1968 being lost in enemy action on that day;
– the Brigade Commander cannot be identified;
– there is no independent record of the rescue on 19 August 1968 of Flt. Lt. Cooper and the Brigade Commander;
– there is no independent record of a pilot being killed;
– there is confusion about the time of day of the incident;
– the location of the crash site is uncertain;
– there is no independent eye witness evidence of the incident nor are there contemporaneous accounts of the incident other than the narratives for a gallantry award;

- *first hand accounts to support the award of the United States' highest gallantry award made at the highest level in the field in Vietnam requiring independent assessment and verification were not prepared at the time; and*
- *the elements of the incident as described for the purposes of a gallantry award are reduced or uncorroborated by subsequent material which has come to light in the intervening years.*

The primary evidential documents available to the Review Panel are the statements of fact made in or in support of the recommendations made by Maj. General Ewell and Lt. Col. Patrick. These statements are supported by the statement of Lt. Colonel James Hoag of the 12th Tactical Fighter Wing.

The Review Panel notes that these documents came to light on 21 November 1974 when Mrs. E. Cooper provided copies to the then Minister for Defence. The Review Panel has been informed that the documents were provided to Flt. Lt. Cooper by an unnamed source during 1974. Copies of the documents are otherwise not available to the Review Panel.

Taking account of all the material made available to it, the Review Panel has found that is unable to be satisfied conclusively that the action took place as described.

Perhaps they were instructed to do this as otherwise the contrary conclusions would have embarrassed the government. A representation was made to Major General Phillips (Rtd.) who was then the National President of the Returned Services League of Australia. He listened to the story and was enthusiastic at the time. However, after he met with the defence minister, Bruce Scott, his eagerness waned considerably. He oversaw the Vietnam End of War List.

If I were General Ewell or one of the numerous brave combat soldiers who have raised the recommendation for the Medal of Honor or the Air Force Cross, I would be extremely insulted by the suggestions made in this panel's findings. War is not a game of chess. Those killed in action are permanently removed from the board on both sides. Battles are bloody, disorganised and disorientating for the participants. The defining battle, but not the only

APPENDIX—THE MEDAL OF HONOR ISSUE

one, for the Australian military in the Vietnam conflict was the Battle of Long Tan that occurred on 18 August 1966. It was a defining moment for the Australian diggers because if they had not stood firm, then the whole of the Australian Task Force at Nui Dat may well have been wiped out. For almost all of the Australian soldiers of 'D' Company 6 Battalion RAR the engagement with the enemy at Long Tan was their first experience of real combat. The defining book of that battle is called, unsurprisingly, *The Battle of Long Tan* as told to Bob Grandin by the people who took part. Bob Grandin was a Huey Pilot with Australia's 9 Squadron in 1966 and took part in the battle. The book highlights shortcomings in the Australian Military Command at the time and also the deviousness in the way Imperial awards for gallantry during the Battle of Long Tan were apportioned.

In the case of my action on the 18 August 1968 I believe I am the only living person who was there! A three-star general with an impeccable military record recommended to the Australian military that I be awarded Australia's highest award. He did this without my knowledge. In the US forces it is customary to make an award for gallantry punctually after the event. Also unknown to me I had been recommended for the Bronze Star for Valor by Lieutenant Colonel Patrick in August 1968. He sent the recommendation for that award to 504 TASG and by them to HQAFV, where it was kept. Therefore I was never awarded it in South Vietnam and nor did I receive any other US awards there that were formally processed. The panel had no need to investigate the details of the events of 1968 that occurred over 30 years ago. That investigation had already been done prior to General Ewell recommending me for the Medal of Honor. There is no mystery surrounding the missing seven eyewitness accounts, it is more a case of some intrigue.

During 2001 the US Air Force Awards Branch contacted the Australian Government asking for details of my Vietnam service as they were once again processing an Air Force Cross on me. The Australian Government may have sent some details but they also sent a copy of the flawed panel findings. They did not send a copy of my rebuttal to the findings where I show that the document is flawed. When the Americans found out that the Australian Government was not going to sanction the award, they ceased processing it.

When I discovered this I made a complaint to the Australian defence minister who claimed that a copy of the panel's findings was not sent to the Americans. Subsequently our defence force ombudsman took up the issue and for several months the defence minister denied they had sent the document to the United States. I provided proof to the ombudsman that the defence minister was incorrect. After several months of denial, the defence minister finally admitted they had sent the document, which prevented the AFC being made because, quote, 'they thought it appropriate!' The ombudsman wrote to the defence department pointing out that, 'it would have been preferable, and more transparent, if Defence had provided a copy of my rebuttal to USAF'.

In the absence of any response from defence, an approach was made to the present veterans' affairs minister, De-Anne Kelly, in 2004. Her reply a few months later did not indicate that she had understood what had been presented to her as she did not address the proposed matter and defended the flawed report. She then went on to assure the reader of her continued support and concern for the Vietnam veterans.

In 1995 the Australian Government changed its policy on foreign awards and allowed Australians to receive and wear their US awards. An application form was produced under the ANS (Awards and National Symbols Branch) for a serviceman to make a claim for the recognition of any US award received. Of course, the government knew that they were holding most of the documentation needed for the serviceman to complete the application. The government omitted to advise the servicemen whose documentation they had intercepted. Over the years I managed to obtain copies of a lot of the US documentation pertaining to myself. Using the government procedure set down, I made application in 1999 to have my US awards recognised. There was no response to this application and the documentation I sent

APPENDIX—THE MEDAL OF HONOR ISSUE

disappeared. I have not heard of anyone who has been successful using this government-inspired procedure. To me it appears the government were merely trying to round up any extraneous US awards that they had not yet intercepted, thereby removing further evidence of US awards to Australians in Vietnam.

The defence minister has stated that it is not protocol for the Australian Government to influence a foreign government in awarding an Australian one of its medals. The question then arises why the defence minister considers it protocol to influence a foreign government into *not* making an award as the defence department has done in my case, twice.

The panel has disregarded the statements of the living officers who recommended the Medal of Honor on me. That is, Lieutenant General Ewell and Colonels Patrick, Walker, Hoag, Wright, Garvin and Gately. It is of interest that, when General Ewell was commander of II Field Force, the Australian Brigade was under his operational control. The defence department is in effect distrusting the word of one of its supreme commanders. The panel has further disregarded the statements of Air Vice Marshal Reed, General Hughes, General Emerson, Colonel Nelson, and warrant officers Benson and Thompson. In preference, the panel has based its decision on the absence of the documents they knew were disposed of by the defence department. With this they were able to stamp 'FILE CLOSED' on my case.

Mr Ron Horton of the National Symbols Office told Keith Payne VC, that if further evidence were to come to light they would reopen the case. However, when further evidence is presented the defence minister quickly announces, 'FILE CLOSED', the result they set out to achieve.

Subsequent approaches to the present Minister for Veterans' Affairs, De-Anne Kelly, have met with similar ignorance. I have approached the minister on several occasions regarding the Air Medals which all FACs should have received. The minister is shielded by a team of minders who give the matter the mirror treatment: 'We are looking into it!' However, they do nothing and never return phone calls or answer letters. Recently 15 Air Medals were approved by the US for Australian FACs. These recommendations, amounting to some 700 pages, were forwarded by the US Embassy in Canberra to the Australian Awards and National Symbols Office (ANSO), which should have processed and forwarded them to the department of defence for concurrence. The ANSO claim they have no knowledge of, or have misplaced, this rather large volume of paperwork. These are 15 of the 430 US Awards to Vietnam Veterans which have not reached the recipients due to Australian Government intervention.

This tardy treatment was not afforded to John Howard when he was recently presented with a high-profile US Award.

This whole matter has been brought to the attention of Major General Bill Crews at the National Level of the RSL. However, his political persuasion prevents him from giving support to the hundreds of veterans who have been affected.

A leading political journalist once said that, 'The government will never hold an enquiry unless they know beforehand what the results will be.'

APPENDIX—THE MEDAL OF HONOR ISSUE

Document A – General Ewell's letter to Bronwyn Bishop

LT. GENERAL JULIAN J. EWELL, USA.,RETD..
9024 BELVOIR WOODS PARKWAY
FORT BELVOIR
VIRGINIA 22060
UNITED STATES OF AMERICA

The Hon. Bronwyn Bishop.
Minister for Defense Industry, Science & Personnel
Parliament House
Canberra
ACT 2600
Australia

20 May 1997

> RECEIVED
>
> 2 3 JUN 1997
>
> MINISTER FOR DEFENCE
> INDUSTRY, SCIENCE & PERSONNEL

Dear Minister Bishop,

On 20 August 1968 I made a recommendation for the Congressional Medal of Honor to be awarded to an Australian, Flight Lieutenant Garry G. Cooper, which appears to have been forwarded to the Australian system and never returned to the US authorities for processing.

The US cannot award the Medal of Honor to Cooper as he is not a US citizen.

As the originating officer of this recommendation I would strongly support Australia showing its highest recognition and awarding Cooper the Victoria Cross.

Cooper's actions of August 1968 brought great credit to himself and in the best traditions of Australian military personnel.

Please find enclosed a copy of the original recommendation and eye witness report.

The gallantry of Cooper, the events and circumstances of August 1968 are as vivid to me today as they were nearly 30 years ago.

Advice as to your actions will be appreciated

Yours sincerely

Julian J. Ewell
LTGEN. USA. RETD

SOCK IT TO 'EM BABY

Document B – General Ewell's Recommendation
August 20 1968

DEPARTMENT OF THE ARMY
HEADQUARTERS 9TH INFANTRY DIVISION
APO San Francisco 96370

006155

AVDE-CG

20 August 1968

THRU: Commanding General
7th USAF
Bien Hoa
APO 96307

TO: Lieutenant Colonel James T. Patrick, USAF
Commanding Officer
19th Tactical Support Squadron
Bien Hoa
APO 96227

Dear Colonel Patrick:

I recommend the immediate award of the CONGRESSIONAL MEDAL OF HONOR to Flight Lieutenant Gary G. Cooper, 0219964, Royal Australian Air Force, one of the men in your command serving with the Ninth Division.

Throughout his service as a FAC he has made himself a legend and is highly respected by all my officers. The narrative description of his gallant actions on 18 August 1968 is attached. Although untrained in ground combat he displayed great coolness and heroism under completely overwhelming odds and saved one of my most valuable Brigade Commanders. His actions would have been highly commendable even if carried out by a veteran in ground combat.

Air Force regulations do not permit me to make an award so I strongly recommend the highest recognition.

With warm regards,

Sincerely,

[signature]
JULIAN J. EWELL
Major General, USA
Commanding

1 Incl
as

The Brigade Commander metioned above was Colonel Robert E. Archer, Commanding Officer, 2nd Brigade, 9th Infantry Division.

[signature]
JULIAN J. EWELL

APPENDIX—THE MEDAL OF HONOR ISSUE

Document C – Letter from Group Captain Blyth

Director of Honours, Employment and
Administrative Policy

12 Sep 97

Personnel Executive

AD – HFA

PETER COOPER - INDEPENDENT REVIEW.

1. Further to our discussions, following further contemplations, I am convinced that we need to act to ensure that the review of this case is seen to be "objective and independant".

2. There needs to be a review panel. The chairperson should not be the Force and should not be part of DPE. Suggest a SES(E). The next panel member should be a senior officer (POLBorA) from out of Defence. Preferably DAS. You should be the third member whose role is to present the material and guide the deliberations.

3. Please give it some thought and consider the various 'how to' aspects. If we don't do it properly we will only have to do it again!

GPCAPT A.L. BLYTH, D-4-23, Ext 52372

Document D – Terms of Reference

Flt. Lt. G. G. Cooper on 18/19 August and any subsequent recommendation for an imperial gallantry award.
1. To investigate and report on an incident involving Flt. Lt. G. G. Cooper, RAAF, which occurred on 18/19 August 1968, in Vietnam, while he was attached for duty with the US Air Force.
2. Without limiting scope and generality of the review, the Panel is to seek evidence and report on the following matters:
   a. Ascertain if the action of Flt. Lt. G. G. Cooper, on 18/19 August 1968, took place as described;
   b. Establish whether the action of Flt. Lt. G. G. Cooper on 18/19 August 1968, was worthy of a recommendation for the award of the Victoria Cross or for another Imperial gallantry award;
   c. Make a recommendation of what, if any, formal recognition Flt. Lt. G. G. Cooper should receive for his action on 18/19 August 1968;
   d. Ascertain if the US Air Force made a recommendation to the Australian Forces in Vietnam, at any time, that Flt. Lt. G. G. Cooper be recommended for an Imperial gallantry award for the action on 18/19 August 1968 and;
   e. The impact that any recognition provided at this time for the action of Flt. Lt. G. G. Cooper would have on the End of War List Vietnam deliberations.

Document E – Lieutenant General Julian J. Ewell, United States Army (ret.)

Lieutenant General Julian J. Ewell is a graduate of the US Military Academy at West Point and follows on from two generations of general officers in the U.S. Army. After receiving his infantry commission, General Ewell took his first assignment with the 29th Infantry at Fort Benning, Georgia. He was part of the initial group

APPENDIX—THE MEDAL OF HONOR ISSUE

to enter the 501st Parachute Infantry, which was a new type of service within the US Forces. He was a part of the development team introducing this new concept to the US Army.

During the early stages of the Second World War, General Ewell served with various parachute units as a Master Jump Instructor before taking command, as a full Colonel, of the 501st Parachute Infantry in 1942. He remained in command of this unit until severely wounded in action during the siege of Bastogne. The 501st participated in European campaigns as part of the 101st Airborne (Screaming Eagles) Division. General Ewell jumped into Normandy on D-Day and into Holland in the autumn of 1944.

General Ewell returned to the US after the Second World War as a student, and later he became an instructor at the Command and General Staff College, Fort Leavenworth, Kansas. After this assignment, General Ewell was assigned to Berlin as executive officer to the US Commander, Berlin, and subsequently became Chief Planner of the Seventh Army at Stuttgart, Germany. He then attended the Army War College at Carlisle Barracks, Pennsylvania, before being posted to Korea in 1952 as commander of the 9th "Manchu" Infantry Regiment of the Eighth Army. On returning to the US, General Ewell spent four years at West Point as commander of a cadet regiment and later as assistant commander of cadets.

In 1958–59, he attended the National War College and on graduation served on the Army General Staff as a planner for two years. General Ewell was transferred to the White House as Executive Assistant to the Military Representative of the President, General Maxwell D. Taylor, to whom General Ewell had given jump qualification during the Second World War. He then moved to the Pentagon as executive to the Chairman of the Joint Chiefs of Staff. In the spring of 1963, he went to Germany as Assistant Division Commander of the 8th Infantry Division and was transferred in June 1965 to Frankfurt as Chief of Staff, V Corps, before assuming command of Fort Belvoir, Washington in June 1966.

He then became Deputy Commander and Chief of Staff for the Combat Developments Command. In 1968 General Ewell took command of the 9th Infantry Division in Vietnam, followed by the command of III Field Corps where he was number two to General Westmoreland. This command was followed as Military Chairman of the Paris Peace Talks and several senior military assignments before his retirement.

General Ewell holds numerous U.S. and foreign decorations, including the Distinguished Service Cross, Distinguished Service Medal with Oak Leaf Cluster, Silver Star with Oak Leaf Cluster, Legion of Merit with Oak Leaf Cluster, Bronze Star, Air Medal with Four Oak Leaf Clusters, Purple Heart, Combat Infantryman's Badge and Master Parachutist's Badge.

Document F – Citation to accompany the award of the Bronze Star Medal to Garry G. Cooper

Flight Lieutenant Garry G. Cooper distinguished himself by heroism while participating in ground combat near Cai Be, Republic of Vietnam, on 18 August 1968. On that date, just before dusk, Flight Lieutenant Cooper was acting as ALO on the Command helicopter during an enemy engagement when his pilot was shot dead at low altitude. Flight Lieutenant Cooper was in front of the enemy position. Flight Lieutenant Cooper was able to reduce the helicopter's ground impact only 200 meters in front of the enemy position. Under a hail of automatic weapons fire he assisted the injured Colonel from the wreckage to a nearby ditch. Throughout the night Flight Lieutenant Cooper fought off several enemy attacks killing at least ten of the enemy at close quarters. The helicopter pickup at dawn was carried out under intense fire and Flight Lieutenant Cooper killed a further two enemy with his empty handgun as they tried to overpower him. The outstanding heroism and selfless devotion to duty displayed by Flight Lieutenant Cooper reflects great credit upon himself and the Free World Military Forces.

APPENDIX—THE MEDAL OF HONOR ISSUE

Document G – Missing Eyewitness Documents

DPSA

Decorations --- Flt Lt Garry G. Cooper, 0219964, RAAF

Cathay Pacific Airways Ltd
Tan Son Nhut, Saigon

1.   Attached decorations pertaining to Flt Lt Garry G. Cooper, 0219964, RAAF, are forwarded for your information and necessary action.

2.   The Bronze Star Medal is an additional award while the other documents are duplicates not required by this office.

3.   Last known forwarding address for Flt Lt Cooper was in care of Cathay Pacific Airways Ltd.

FOR THE COMMANDER

JOHN E. DUNNING, Captain, USAF
Chief, Awards Branch
Personal Affairs Division

14 Atch
1.   Special Order
2.   BSM
3.   SS 1st OLC
4.   DFC 3rd OLC
5.   Vietnam Cross of Gallantry with translation
6.   7 statements
7.   AF 77a

[Author note: In the attachments listed, item 6 is the one relating to the seven eyewitness statements regarding the events of the 18 August 1968 that are missing from my file.]

SOCK IT TO 'EM BABY

Document H – Consider for an Upgrade by Lt Colonel Patrick

FOR OFFICIAL USE ONLY   N.B. - Consider for upgrade.

RECOMMENDATION FOR DECORATION	DATE 23 August 1968
TO: (Organization and address) 504 TASG APO 96227	FROM: (Organization and address) 19 TASS APO 96227

RECOMMENDATION

RECOMMEND INDIVIDUAL INDICATED BE AWARDED

1. NAME OF DECORATION (Indicate number of clusters, if appropriate)
   Bronze Star for Valor

2. RECOMMENDATION IS BASED ON:	3. INCLUSIVE DATE(S) OF ACT, ACHIEVEMENT OF SERVICE	
[X] HEROISM [ ] MERITORIOUS SERVICE (Based on completed period of service) [ ] OUTSTANDING ACHIEVMENT	FROM 18 August 1968	TO 19 August 1968

PERSONAL DATA ON INDIVIDUAL BEING RECOMMENDED

4. LAST NAME - FIRST NAME - MIDDLE INITIAL Cooper, Garry G.	5. AFSN 0219964	6. GRADE Flight Lieutenant
7. PRESENT ORGANIZATION AND STATION 19 TASS APO 96227	8. PRESENT DUTY ASSIGNMENT Forward Air Controller	
9. PRESENT HOME ADDRESS Officers' Mess, RAAF Base, Williamtown, NSW, 2301, Australia	10. ORGANIZATION OF NEXT DUTY ASSIGNMENT (If applicable) NA	
11. ORGANIZATION, DUTY ASSIGNMENT AND GRADE AT TIME OF ACT OR SERVICE. 19 TASS Forward Air Controller Flight Lieutenant	12. DATE OF PROMOTIOIN TO GRADE IN WHICH SERVING 16 Dec 65	13. INDIVIDUAL'S SERVICE IN AIR FORCE SINCE ACT OR SERVICE HAS BEEN HONORABLE [X] YES [ ] NO

14. DATE OF REASSIGNMENT, RETIREMENT OR SEPARATION, AS APPLICABLE: DEROS: 08 Nov 68
    RETIREMENT OR SEPARATION IS [ ] VOLUNTARY [ ] INNVOLUNTARY, AND THE FOLLOWING SERVICE DATES APPLY:
    CDOS_____ TAFCSD_____ TAFMSD_____
    PLSD_____ TFCSD_____ TMSO_____

15. PREVIOUS UNITED STATES DECORATIONS, COMPLETE AUTHORITY THEREFOR, AND INCLUSIVE DATES OF SERVICE RECOGNNIZED (Do not include service medals, battle credits, unit citations or foreign decorations)
    None

16. ARE OTHER RECOMMENDATIONS FOR AWARDS TO THIS INDIVIDUAL PENDING? (If yes, state awards) [ ] YES [X] NO

17. ARE OTHER INDIVIDUALS BEING RECOMMENDED FOR THE SAME ACT OR SERVICE? [ ] YES [X] NO

18a. IF ANSWER TO ITEM 17 IS YES, ARE THE RECOMMENDATIONS FOR THE OTHER INDIVIDUALS FORWARDED AS PART OF THIS RECOMMENDATION? [ ] YES [ ] NO

b. IF ANSWER TO ITEM 18a IS NO, EXPLAIN REASON FOR DELAY, INCLUDING DATE RECOMMENDATION(S) WILL BE FORWARDED, AND IDENTIFY THE INDIVIDUAL(S) BY GRADE, NAME, SERVICE NUMBER, PRESENT ORGANIZATION AND STATION.
NA

19. HAS A PREVIOUS AWARD BEEN MADE TO THIS INDIVIDUAL FOR THIS ACT OR SERVICE? [ ] YES [X] NO

20. HAVE ALL AVAILABLE RECORDS AND SOURCES OF INFORMATION THAT WOULD HAVE A BEARING ON THIS RECOMMENDATION BEEN CONSIDERED AND NO CONDITION EXISTS WHICH WOULD MAKE APPROVAL OF THIS AWARD INAPPROPRIATE? [X] YES [ ] NO

21. IF AWARD IS POSTHUMOUS, OR INDIVIDUAL RECOMMENDED IS MISSING IN ACTION OR A PRISONER OF WAR, LIST NAME, ADDRESS AND RELATIONSHIP OF NEXT OF KIN. NA

22. DATE WHICH PRESENTATION OF AWARD IS DESIRED IF APPROVED.
    23 September 1968

23. IF APPROVED, FORWARD FOR PRESENTATION TO (Organization and address)
    19 TASS APO 96227

AF FORM 642 NOV 66   PREVIOUS EDITION OF THIS FORM WILL BE USED UNTIL STOCK IS EXHAUSTED.   FOR OFFICIAL USE ONLY

APPENDIX—THE MEDAL OF HONOR ISSUE

24. NARRATIVE DESCRIPTIION (Description of the act, achievement or service, including specific dates, places and facts. If additional space is needed, use plain paper 8 x 101/2 bond paper, the last sheet of which must be signed by the recommending individual.)

Although Flight Lieutenant Garry G. Cooper has been strongly recommended for the Medal of Honor, regulations do not permit foreign nationals to receive this award. Flight Lieutenant Cooper can not be recommended for the second highest award, the Air Force Cross, as his gallantry did not take place in the air. Therefore the maximum I can submit is the Bronze Star Medal for Valor which is totally inadequate and I strongly urge the British to consider Cooper for their highest recognition.

Flight Lieutenant Cooper distinguished himself by gallantry in action in connection with military operations against an opposing armed force as an Air Liaison Officer near Cai Be, Republic of Vietnam on 18 August 1968. On that date, late in the afternoon, Flight Lieutenant Cooper was flying with a pilot and the Brigade Commander on the Command Helicopter. They were flying at 200 feet between the friendly and hostile forces. The helicopter was taking numerous hits from the intense automatic weapons fire but it was imperative that they continue directing operations from this position as the friendly forces were pinned down and taking heavy casualties. After thirty minutes under heavy fire the pilot was shot dead at 200 feet and the helicopter dived toward the ground at tremendous speed. The Brigade Commander had been hit in the back of the neck with the same bullet that disintegrated the pilot's head and ricocheted off Cooper's helmet, stunning him. Although dazed and covered in blood and brain tissue, Cooper managed to reach across for the controls, overpower the dead pilot and reduce what would have been a fatal impact with the ground. The crash was in open rice paddy country only 200 meters in front of the enemy lines and the helicopter became the primary target for their fire. Friendly troops could hear the automatic weapons fire slamming into the wreckage while Cooper, although finding it difficult to move due to a back injury, assisted the Colonel to a near-by dyke. Here they set up a defensive position as they could move no further and the friendly forces were not in a position to help. They were waist deep in water and throughout the night the enemy made several attempts to creep along the dyke to their position. Each time Cooper managed to ward off the attacks and killed at least ten of the enemy at close range. By morning the situation had eased due to constant air strikes on the enemy positions and a pickup helicopter was called for. Weapons fire was again heavy during the pickup and the helicopter had to move away making it necessary for the Colonel and Cooper to run in the open to board it. Now out of ammunition, as Cooper had been covering the Colonel while he was hoisted aboard, he was attacked by two of the enemy who he killed with his empty hand-gun before leaping into the helicopter. Flight Lieutenant Cooper's gallantry and professionalism reflect great credit upon himself, the Royal Australian Air Force and the Free World Military Forces.

25. TYPED NAME, GRADE AND TITLE OF INITIATING INDIVIDUAL	26. SIGNATURE
JAMES T. PATRICK, LT Colonel, USAF Commander	James T Patrick

27. ATTACHMENTS

NUMBER	DESCRIPTION (citation and supporting statements or other official documents)
1	Recommendation for award of Medal of Honor. (Major General Ewell)
2	Narrative in support of recommendation.

SOCK IT TO 'EM BABY

Document I – Colonel Hoag's Letter 19 August 1968

DEPARTMENT OF THE AIR FORCE
HEADQUARTERS 12TH TACTICAL FIGHTER WING (PACAF)
APO SAN FRANCISCO 96326

REPLY TO
ATTN OF: DCO

19 August 1968

SUBJECT: Recommendation for Award

TO: Deputy Director
III DASC
APO 96227

1. Only two days ago we forwarded a Commendation of Forward Air Controller on Tamale 35. Another was sent on July 1968. It is therefore with even greater regret that the 12th TFW witness this outstanding Forward Air Controller crash during combat.

2. This statement is written so that the valorous actions of Tamale 35, Captain Garry Cooper of the Australian Airforce, on 18 August 1968 will not go unrewarded.

3. On that date two of my F-4 pilots were working with Tamale 35 on a heavily defended target at Cai Be, near Rach Kein, when Captain Cooper's helicopter crashed very close to the hostile position. Without FAC direction, the F-4 pilots could not provide close air support and circled helplessly.

4. When last seen, Tamale 35, distinctive by his Australian flight-suit, was half carrying an Infantryman towards an embankment under what must have been highly hazardous conditions. The situation as viewed did not appear survivable.

5. We of the 12th TFW sincerely hope that Tamale 35 will soon return to provide us with his exceptional Forward Air Controlling.

JAMES H HOAG JR.; Colonel, USAF
Deputy Commander for Operations

APPENDIX—THE MEDAL OF HONOR ISSUE

Document J – Colonel Heard's Award Summary in 1978

DEPARTMENT OF THE AIR FORCE
HEADQUARTERS UNITED STATES AIR FORCE
WASHINGTON, D.C.

COPY TO
OF: AF/CVAIA   30 October 1978

SUBJECT: Recommendations for Awards (Major Garry G. Cooper, RAAF)

TO: Hq AFMPC/MPCASA2

1. This office concurs in the attached recommendations (7) for awards to Major Garry G. Cooper, Royal Australian Air Force (RAAF). The recommendations are for action in the Southeast Asia in 1968 and are submitted IAW AFR 900-48 (para 3-7b). The statement (Atch 2, para 2) from USDLO Hong Kong that the awards were apparently forwarded to RAAF authorities, but not presented, appears to be substantiated by Atch 3, para 2b (DEFAIRA Canberra message to the Australian Air Attache in Wash DC).

2. A background investigation has been requested. Concurrence will be obtained from the U.S. Ambassador/USDAO and DIA, and forwarded when received.

3. The following is provided as background information:
   a. In March 1975 a request from the Australian Embassy was received by this office for information on any U.S. decorations Major Garry G. Cooper, RAAF, may have received for actions in SEA. Subsequent investigations revealed a recommendation for the Bronze Star for Valor (later upgraded to the Air Force Cross). This award was reprocessed by the originator and approved, but presentation was deferred because of lack of information; i.e., eyewitness or name of brigade commander whose life Major Cooper allegedly saved.

   b. We have now obtained copies of nine other recommendations for awards which were not received by Major Cooper. These were held in abeyance while further attempts were made to obtain evidence in support of the recommendation for the Air Force Cross. Again, we were unsuccessful; therefore, this office feels the attached recommendations should not be held any longer and, accordingly, they are being forwarded for processing (two were Army awards and have been forwarded to the Department of the Army for action).

FOR THE CHIEF OF STAFF

ALLAN P. HEARD, Colonel, USAF
Chief, International Affairs Division
Office of the Vice Chief of Staff

3 Atch
1. Recommendations (7)
2. USDLO Hong Kong
Msg 1231, DTG 310229Z
JUL 79
3. DEFAIRA Canberra Msg
S201, DTG 150914Z Oct 75

cc: USDLO Hong Kong

Underwrite Your Country – Buy U.S. Savings Bonds

Document K – Biography of Air Vice-Marshal Jim Flemming

Air Vice-Marshal J.H. Flemming AO, DFC, AM (US), FAIM, FCIT, RAAF (Ret.)

James Hilary Flemming trained as a pilot in Australia and Canada under the Empire Air Training Scheme, graduating in 1944 as a Sergeant. In 1946 he was posted to RAAF Williamtown to convert to Mustang fighters and then to No. 78 Squadron of No 78 Wing. He served with the British Commonwealth Occupation Force in Japan, and with No. 77 Squadron in the Korean War. Whilst serving in Korea he was awarded the American Distinguished Flying Cross and the American Air Medal and was also mentioned in despatches. In 1950 he was commissioned as Flying Officer.

On his return to Australia he spent four years as a flying instructor at various training units instructing pilots in Tiger Moths, Wirraways, Vampires and Meteors. He completed the first Fighter Combat Instructors' Course in 1955 and in 1956 he formed and led the RAAF's first official aerobatic team, the Meteorites.

In October 1958, while on exchange duty in America, Jim Flemming became the first member of the RAAF to pilot an aircraft at twice the speed of sound when flying an F104C Starfighter. As his career continued, he led squadrons in Malaysia and Thailand and was appointed Senior Administrative Staff Officer at Headquarters Support Command. In 1973 we was promoted to Air Commodore and appointed as Officer Commanding RAAF Williamtown. In 1975, while attending the Royal College of Defence Studies in London, he was named in the initial listing of the new Australian Honours as a Member of the Order of Australia.

After more high-level appointments, in January 1978 Flemming was promoted to Air Vice-Marshal and provided by the Australian Government to be Commander of the Five Nation Integrated Air Defence System of Malaysia and Singapore. In 1982 he returned to Canberra to become Chief of Air Force Operations in the Department of Defence and in the Queen's Birthday Honours he was elevated to be an Officer of the Order of Australia, the first person to be so honoured since the inception of the Australian honours system.

After 40 years of continuous service, Jim Flemming resigned from the RAAF and was appointed a Director of the Australian War Memorial, which he held until 1987. He still owns a civil commercial pilot's licence and is the owner of a Cessna XP11 Hawk aircraft. The Air Vice-Marshal is Chairman of the Arts and Museums Advisory Board in the Northern Territory and has been instrumental in the setting up and successful operation of the Aviation Heritage Centre in the Territory.

APPENDIX—THE MEDAL OF HONOR ISSUE

Document L – Enemy contact area, 18 August 1968

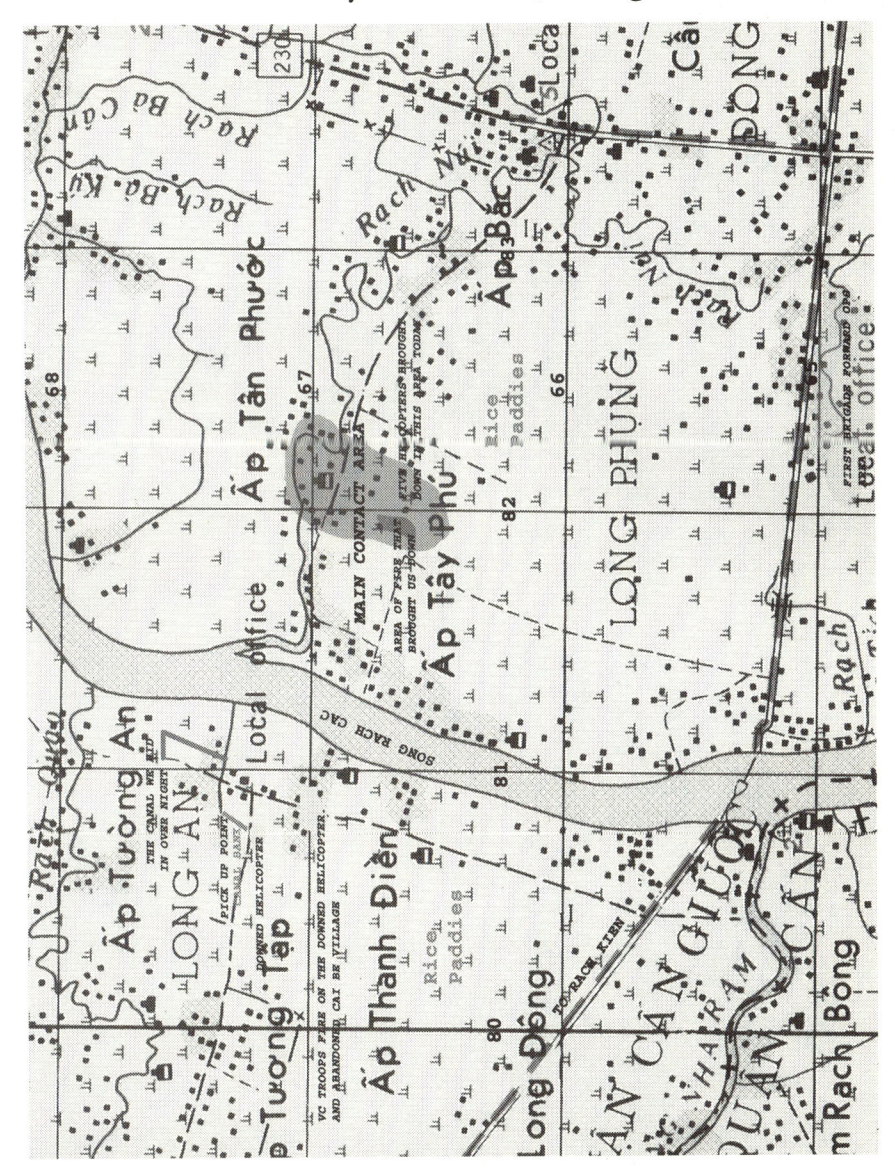

303

GLOSSARY

A1	Douglas Skyraider propeller-driven single-engine strike aircraft
A4	McDonnell Douglas Skyhawk single-engine strike aircraft
A37	Cessna Dragonfly jet aircraft developed from the T37 for counterinsurgency operations
AC47	Douglas C47 transport converted into a gunship by adding General Electric SUU-IIA miniguns
AFC	Air Force Cross—an Australian non-combat award and the second highest US award for bravery
AFHQ	air force headquarters
AFTS	Advanced Flying Training School
AFV	Australian Forces Vietnam
AGCP	air ground control party
AGL	above ground level
AHC	assault helicopter company
AK47	Soviet Avtomat Kalashnikova automatic 7.62 mm assault rifle used by the Viet Cong
ALO	air liaison officer
AM	air medal
ANARE	Australian National Antarctic Research Expedition
AO	area of operation
AOG	aircraft operationally grounded
APC	armoured personnel carrier
AR15	basically an M16 with shortened barrel and folding stock
arty	artillery
ARVN	Army of the Republic of Vietnam
Asap/ASAP	as soon as possible
ATC	air traffic control
ANSO	awards and national symbols office
AWOL	absent without leave

GLOSSARY

B52	Boeing Stratofortress used by the Strategic Air Command, USAF
B57	Martin Canberra strike twin-jet aircraft
Battalion	US Army unit consisting of three or more companies and 600 combat soldiers
BDA	bomb damage assessment
BFTS	Basic Flying Training School
Birddog	Cessna O-1 FAC aircraft
'Bobcat'	radio call-sign of F100 fighters
BOQ	Bachelor Officers' Quarters
Brigade	US Army unit consisting of three or more battalions
btry	artillery battery
bunker	Any dug-out or fighting trench
'Buzzard'	radio call-sign of F100 fighters
C46	Curtis Commando transport aircraft
C119	Fairchild Flying Boxcar twin—boom transport aircraft
C123	Fairchild Provider cargo and defoliant aircraft
C130	Lockheed Hercules transport aircraft
'Caribou'	C7 twin engine STOL transport
CASA	Civil Aviation Safety Authority
CAR15	submachine gun version of the M16 rifle with a shortened barrel and folding stock
CBU	cluster bomb unit
CDR	commander
CFI	chief flying instructor
CFS	Central Flying School
'Charlie'	nickname for the VC
chieu hois	VC defectors to the South Vietnam Government
Chinook	CH47 twin engine transport helicopter with rotors fore and aft
CIP	combat instructor pilot
CO	commanding officer
Company	US Army unit usually consisting of 150 combat troops

cong moui	mosquitoes
III Corps	military region across Vietnam centred around Saigon
COSVN	Central Office South Vietnam
crew chief	ground maintenance person in charge
CTZ	corps tactical zone
DASC	direct air support centre
DEROS	date of eligible return from overseas (end of tour)
DFC	Distinguished Flying Cross
Division	US Army unit consisting of two or more brigades and 10 000 men
DRV	Democratic Republic of Vietnam
DSO	Distinguished Service Order
dust-off	medical evacuation usually by helicopter
F4	McDonnell Phantom fighter strike jet aircraft
F86	North American Sabre jet fighter
F86D	North American Sabre jet fighter (all weather version)
F100	North American Super Sabre strike jet aircraft
FAC	forward air controller
FACing	forward air controlling
FDC	fire direction centre
FDO	Fire Direction Officer
flak	bursting shells fired by anti-aircraft guns
flare ship	any aircraft or helicopter that drops illumination flares
FM	frequency modulation—for air-to-ground radio communications
FNG	'fucking new guy'—derogatory term for inexperienced soldier
FO	forward observer
FOB	forward operating base
FOD	foreign object damage
FOI	Freedom Of Information
freq	frequency
FSB	fire support base
fox mike	spoken term indicating FM

GLOSSARY

free-fire zone an area completely under enemy control permitting unlimited fire power against anyone or anything in the zone
friendlies troops fighting against the VC or NVA

G measure of gravitational pull of an aircraft manoeuvring—a pilot pulling 4G is incurring forces on his body four times his own weight
GCA ground controlled approach—a method of controlling an aircraft by radar to line it up with the landing runway under instrument conditions
GSA Gibbes Sepik Airways
Gooks general term used when referring to the enemy
Grunt nickname for US foot soldier, due to his occupation causing a lot of time wallowing in mud
gunship any aircraft or helicopter armed for tactical operations

'Hammer' fighter call-sign
helipad helicopter landing area
helo helicopter
high-drag bombs equipped with fins that open on release to slow their descent
hot insertion landing soldiers by helicopter into an area of operation under enemy fire
hooch basically, any structure to house humans or animals. It is a derivative from the Japanese word 'uchi', meaning house or home.
HQ headquarters
HQAFV Headquarters Australian Forces Vietnam
Huey see UH-1
Hun nickname for F100

in-country within the conflict area of South Vietnam
in hot fighter rolling in on a target and ready to fire
insertion landing soldiers in the area of operation by helicopter

IP	initial point—a well-defined point for the fighters to commence their attack sequence
JOC	joint operations centre
JP4	jet aircraft fuel (kerosene)
KBA	killed by bomb action
KIA	killed in action
kg	kilogram
klick	military jargon for approximately 1000 metres or .6 of a mile.
klm	kilometre
Kph	kilometres per hour
LZ	landing zone for helicopters
LOH	light observation helicopter
m	metre
M16	5.56 mm assault rifle
MACV	military assistance command vietnam
Magpie	radio call-sign for the Australian Canberra bomber
Marston Matting	see 'PSP'
max ord	maximum height of artillery fire
medivac	medical evacuation, usually by helicopter
METO	maximum except for takeoff
MIA	missing in action
mike-mike	spoken term indicating millimetre, usually denoting a weapon's calibre, eg. 20 mike-mike = 20 mm cannon
Mk82	streamlined bomb set to detonate on ground contact or with a delay for penetration
Moonshine	radio call-sign for flare dropping AC47
mm	millimetre
MRF	Mobile Riverine Force

GLOSSARY

napalm	petroleum jelly fire bomb
nipa palm	heavy sago type palm generally along river banks and swamps
NLF	National Liberation Front
NVA	North Vietnamese Army
O-1	Cessna 'Birddog' FAC aircraft
OH23	Hillier 'Raven' observation helicopter
OV10	North American Rockwell 'Bronco' FAC aircraft
OC	officer commanding
O-Club	officers' club
OCU	Operational Conversion Unit
opcon	under operational control
PAVN	People's Army of Vietnam
PFC	private first class
PLAF	Peoples' Liberation Armed Forces
POW	prisoner of war
PRO	public relations office
PSP	pierced steel planking (or Marston Matting) that is laid across mud surfaces
PTSD	post traumatic stress disorder
Purple Heart	medal awarded to US troops killed or wounded in action
RAAF	Royal Australian Air Force
ramp	area on an airport where aircraft are parked
RAR	Royal Australian Regiment
RF101	McDonnell Douglas Voodoo reconnaissance jet aircraft
R & R	rest and recreation (leave) or rest and recuperation (wounds)
recce	reconnaissance
recon	reconnaissance
REMF	'rear echelon motherfucker'—derogatory term used for non-combatants

revetment	high, thick walls built around parked aircraft for protection
RIF	reconnaissance in force
Roger	radio term in the affirmative
RP	rocket projectile
RPG	rocket propelled grenade
RSL	Returned and Services League
Rtd	retired
RVN	Republic of Vietnam
SAN	School of Air Navigation
scramble	takeoff without delay
secondary	a delayed explosion on the ground triggered by weapons detonation
short round	inadvertent or accidental delivery of ordnance sometimes resulting in death or injury to friendly troops
sitrep	situation report
Skyraider	see A1 aircraft
slick	a troop carrying helicopter as well as a low-drag bomb
sortie	the time between takeoff and landing of an aircraft
SP	service police
Spooky	call-sign of the AC47, flare dropping, mini-gun equipped aircraft
stall	when the aircraft wings no longer keep an aircraft flying
Starlight Scope	an image intensifying scope or binocular using light from the moon or stars to give a target image
Super Sabre	see F100 aircraft
SVN	South Vietnam
TACAIR	a term used to encompass all aircraft delivering weapons, literally 'Tactical Air' (support)
TACAN	tactical air navigation aid/beacon
TACP	tactical air control party
TASG	tactical air support group

GLOSSARY

Tamale	radio call-sign for the 9th Division FACs
TASS	tactical air support squadron
Tet	the lunar holiday observed by Vietnam
TFW	tactical fighter wing
TOC	tactical operations centre
UH1	Bell Aircraft 'Huey' helicopter
UHF	ultra high frequency radio
Uniform	spoken term for UHF
USAF	United States Air Force
US Army	United States Army
USO	United Servicemen's Organisation
VC	Viet Cong
Victor	spoken term for VHF
VHF	very high frequency radio
VNAF	Vietnamese Air Force
VR	visual reconnaissance
VT	variable time (fusing)
WIA	wounded in action
Willy Pete	nickname for the white phosphorus smoke 2.75 inch rockets launched from the FAC O-1s to mark targets
wind shear	a condition produced by mixing winds at different heights
WO	warrant officer
WOD	warrant officer disciplinary
WRAAF	Women's Royal Australian Air Force
XO	executive office
Y bridge	vitally strategic bridge joining Saigon to the Mekong Delta and the enemy objective during the Tet and the May Offensive

BIBLIOGRAPHY

Buttinger, Joseph, *Vietnam: A Political History*, Andre Deutsch, 1969.

Davidson, Phillip B., *Vietnam At War—The History 1946–1975*, Oxford University Press, 1985.

Grandin, Bob, *The Battle of Long Tan—As told by the commanders to Bob Grandin*, Allen and Unwin, 2004.

Hannaford, Helen, *Webster's New World Dictionary of the Vietnam War*, Leepson, Marc and Simon and Schuster Inc., 1999.

Truong Nhu Tang, *A Viet Cong Memoir—An Inside Account of the Vietnam War and its Aftermath*, Harcourt Brace Jovanovich Publishers, 1985.

INDEX

Adelaide, 24
Advanced Flying Training School, 34
Agent Orange, 183
Air Vietnam, 254
Alley, Lieutenant Lee, 243, 256
An Nhut Tan, 121
Anderson, Captain Andy, 66, 71, 101, 108, 109, 112, 114, 126, 130, 139, 143, 184, 233
Ansett Airlines, 254–6
Antila, Lieutenant Colonel Eric F., 93, 95, 96, 100
Ap An Hoa, 181
Ap Bac, 234
Ap Tay Phu, 203
Applied Flying Training School, 39
Archer, Colonel, 189, 194, 195, 196, 197, 198, 200, 201, 202, 203, 204, 209–17, 219, 255, 261, 267, 268, 273
Armed Forces TV, 75
Ashworth, Squadron Leader Norm, 43, 45
Astbury, Geoff, 40
Australia National Antarctic Research Expedition (ANARE), 43, 46

Balas, Tex, 243
Ballina, 23
Bankstown, 43, 45
Barnard, Lance, 262, 264, 267, 282
Bartlett, Squadron Leader, 45
Batchelor, Squadron Leader John, 46, 47
Bates, Brian, 277, 279, 279
Battye, Alister, 47
Bearcat aviation base, 62

Beatles, The, 74
Behan, William G., 94
Ben Luc bridge, 71, 88, 92, 131, 132
    The Bowling Green, 92, 130, 135, 151
Ben Tranh, 176
Benson, Colonel, 134, 136, 146, 192
Bien Hoa, 18, 55, 62, 65, 71, 93, 107, 117, 118, 121, 128, 129, 133, 136, 142, 158, 159, 174, 182, 184, 191, 234, 238, 239, 240, 244, 246, 248
Binh Chanh, 137
Binh Phuoc, 76, 77, 79, 80, 128, 134, 148, 195, 196
Binh Thuy, 8, 175
Binh Truong Tay, 135
Bishop, Bronwyn, 259
Blyth, Group Captain A.L., 259, 262
Botesch, Captain Dick, 181
Bower, Jeff, 208
Bowman, Sergeant Harold E., 180
Bryant, Pilot Officer Jock, 49, 59
Buchannan, Dennis, 35
Butler, Pilot Officer Ray, 49

Ca Duoc, 172
Cage, Major Chuck, 233, 238
Cai Be, 165, 173, 178, 179, 190, 197, 198, 199
Cai Lay, 163, 176
Cai Nuoc, 173
Cambodia, 101, 124, 151, 234, 236
Camp Robert Rethune, 76, 79
Can Giuoc, 72, 80, 137, 159, 194, 199, 203
Can Tho, 174, 175, 185
Cannon, Wing Commander Vic, 49

Cargolux Airlines, 254
Carrol, Tom, 231
Cat Lai, 231
Cathay Pacific Airways, 116, 222, 253, 257, 267
Central Flying School (CFS), 45
Champion, Dave, 40
Chiang Kai Shek International Airport, 221
Clark, Bill, 36
Con Son Island, 121
Cooper, Carl, 250
Cooper, Don, 221
Cooper, Garry
  Air Force Cross, 258, 261, 285
  Air Medal for Valor recommendation, 169
  ALO duties, 228–33
  Ansett Airlines, 254–6
  Australian National Antarctic Research Expedition (ANARE), 43–5, 46–7
  awards ceremonies, 116–17
  Bearcat aviation base, 63–5
  British Distinguished Flying Cross, 252
  Bronze Star citation, 257, 261, 262, 264, 282, 285
  call sign 'Tamale 35', 67
  Cathay Pacific Airways, employment at, 253
  Cessna 0-1E 'Birddog', history of, 121–2
  childhood, 24–5
  Commercial Pilot Licence Test, 28
  Cross of Gallantry with Silver Star, 227
  DFC recommendation, 155
  Dong Tam posting, 160
  fighter pilot training, 48–50
  flight instruction, 27
  Gobel Trophy, 42
  helicopter crash, 18 August 1968, 204–15
  Honour Medal First Class, 227
  Joint Warfare Course, 53
  Leaving Honours Certificate, 26
  Letter of Commendation, 190
  lucky charm, 129, 218
  May Offensive and, 76–86
  Medal of Honor recommendation, 189, 218–19, 257–301
  New Guinea, working in, 31–4
  Phan Rang airbase, 57
  Purple Heart, 216, 217
  RAAF training, 35–42
  75 Squadron posting, 51
  Silver Star award, 130
  Tan An posting, 65
  US Distinguished Flying Cross, recommendation for, 116
  Victoria Cross of Australia, 259
  Vietnam FAC posting, 54
  Wings Test, 41
Cooper, Jean, 100
Cooper, Mark, 1, 23
Cooper, Mrs E., 262, 268
Cottrell, Flying Officer Macaulay 'Mac', 54, 55, 59, 62, 238, 248, 250, 251
Cover, Sergeant Mike, 101, 123, 126, 130, 169, 176, 184, 193, 195, 241
Craig, Captain James B., 111
Creer, Brian, 27
Creswell, Dick, 34
Crews, Major General Bill, 288
Cuming, Group Captain, 52
Cutler, Sir Roden, 252

Da Bien Canal, 171
Darwin, 249, 250
De Luca, Lieutenant Colonel Anthony, 87, 88
Deichelman, Sam, 228, 234

INDEX

Deister, Jim, 208
Di Di Mau, 232–3
Djakarta, 116
Dong Tam, 109, 112, 159, 160, 161, 163, 165, 166, 167, 170, 171, 172, 173, 174, 175, 178, 180, 181, 182, 184, 186, 187, 191, 199, 218, 225, 227, 228, 234, 235, 239, 244, 267
Dowling, Air Commodore, 130, 145, 146
Duggan, Terry, 40, 41

East Sale, Victoria, 42
Eberbach, Eric, 28
Ellis, Jack, 50
Ellis, Reg, 27
Emerson, Colonel Hank 'Gunslinger', 201, 202, 210, 271, 276, 277
Ewell, Major General, 130, 171, 187, 188, 237, 257, 258, 259, 262, 263, 267, 277, 279, 282, 285, 287

Feiss, Mick, 50
Flemming, Wing Commander Jim, 51, 56
Flores, Richard J., 94
Forrester, First Lieutenant William L., 81, 82, 85, 87
Fort Walton Beach, Florida, 121
Fox, Major Amos, 57, 59
Franks, Lieutenant Tommy, 96, 100
Frecker, Sergeant Ron, 43

Gibbes, Bobby, 30, 32, 34, 35
Gibbes Sepik Airways (GSA), 30, 34, 35
Go Cong, 245
Goetz, Don, 201
Gogerley, Squadron Leader Bruce, 49
Goroka, 31
Grandin, Bob, 285
Gray, Robin, 34

Gray, Warrant Officer George C., 165, 166, 173, 192
Ground Controlled Approach School, 43

Hackworth, Colonel David, 188
Hai Muroi Tam Canal, 237
Hanrahan, Penny, 1, 21, 22
Harris, Warrant Officer 'Ming', 37, 38
Hau Giang River, 175
Hay, Major General R.A., 130
Hayes, Flight Lieutenant Max, 39
Heath, John, 279
Heidemann, Heidi, 240
Hemphill, Lieutenant Colonel, 173, 192, 228, 229, 230
Ho Chi Minh city, 255
Ho Chi Minh Trail, 92, 124, 135
Hoag, Colonel, 266, 268, 287
Hong Kong, 1, 116
  Kai Tak Airport, 1
Honolulu, 110, 165
Hope, Bob, 74
Horsman, Squadron Wing Commander Bill, 54
Hubble, Group Captain John, 162, 163, 249
Hunt, Colonel Ira, 201

Iverson, Captain Craig R., 178

Johnson, Captain, 62
Johnson, Chaplain James D., 231
Johnson, Jake, 72, 117, 119, 120, 153, 155, 161, 177, 179, 241
Jones, Mister (Warrant Officer), 195, 196, 198, 199, 200, 209
Jones, Ron, 1, 23

Karpys, Tony, 51
Kelly, De-Anne, 286, 287
Khan, Lieutenant Colonel Colin, 250

Kheim Ich, 184
Kichenside, Squadron Leader Jim, 49
Kinh Bac Dong, 151
Kinh Bo Bo Canal, 92, 101, 135, 136, 140, 151
Kinh Doi Canal, 71, 90, 93, 105, 110
Kisling, Sergeant Alan, 100
Kisling, Sylvia, 100
Kokoda Trail, 31
Kowloon Bay, 1

Labuan, 449
Lae, 31
Lai Khe, 277
Lake Tonle Sap, 234
Latham, Captain James, 173
Laurent, Pauline, 88
Lawnton, Tina, 30
Long Binh, 71, 93, 108
Long Binh Hospital, 109
Long Phu Tay, 80
Long Tan, Battle of, 170, 219, 285
Long Thanh North (Bearcat), 62, 131
Ly, Major Bung, 121, 122

MacDonald, Major General E.L., 130
MacDonald, Wing Commander G.G., 260, 262
McGhee, K.J., 260
McNulty, Captain Richard C., 178
Madang, 34
Madang Air Services (MAS), 34
Malaysia, 50
Manser, Peter, 33
May Offensive, 71
Medal of Honor, actions recommended for, 204–15
Mekong Delta, 9, 160, 161
Mekong River, 160, 199
Melbourne, 47
Midway, 122
Mink, Captain Larry, 226, 228, 245

Mobile Riverine Force (MRF), 160, 161, 183
Moc Hoa, 235
Monaghan, Lieutenant Bill, 48
Munn, Ray, 26
Muoi Hai Canal, 234, 235
My Phuoc Tay, 176
My Tho, 160, 181, 182
My Tho Canal, 236, 237
My Tho River, 160, 192, 234, 245

Nakhon Phanom, 49
National Cash Register Company, 29, 30
Nelson, Major Richard 'Dick', 3, 4, 12, 13, 21, 167, 169, 170, 171, 172, 174, 181, 182, 184, 218, 226, 227, 228, 240, 244, 245
New Guinea, 31–35, 131
Newton John, Olivia, 129
Ngang Canal, 151
Nguyen, General Viet Thanh, 227
Nha Be River, 182
North Vietnamese Army (NVA), 4, 70, 81, 92, 97, 101, 124, 135, 136, 140, 142
Nui Dat, 285
Nuvolini, Captain Joe, 233, 236, 238, 239, 244

O'Brien, Brigadier K.J., 260, 263
Olsen, Captain, 61, 62
O'Neill, Paul, 260, 263

Parrot's Beak, 92, 101, 124, 135, 136, 235
Patrick, Lieutenant Colonel James T., 56, 136, 137, 146, 247, 248, 258, 261, 268, 277, 282, 285, 287
Patterson, Flying Officer Pat, 39
Payne, Captain Ike, 66, 67, 68, 108
Payne, Squadron Leader, 45

INDEX

Pearce, West Australia, 39, 40
Peit'ou, 223
Penang Island, 50
Peterson, Captain, 58, 60, 61
Phan Rang Airbase, 57, 61
Phillips, Major General, 284
Phu Tan, 231
Pitts, Colonel John E., 116
Point Addis, 36
Point Cook, 35, 36, 40, 43, 47
Port Moresby, 31
post traumatic stress disorder (PTSD), 39, 256
Price, Fred, 38
Pyman, Flight Lieutenant John, 49

Querry, Sergeant Howard, 88

Rach Cac River, 200, 212, 271, 276, 277
Rach Gieu, 124
Rach Kien, 72, 144, 146, 151, 199, 200
Ragh Xom River, 90
Rahng Sak, 182, 183
Richardson, Sergeant Alan, 43, 46, 47
Robe, 24
Robertson, Captain, 62, 64, 68
Rogers, Dave, 50
Rohe, David, 201
Royal Australian Air Force (RAAF), 34, 35
  Butterworth, Malaysia, 49, 50, 249
  Laverton, 250
  Williamtown, 48, 53, 250, 253
Royal Australian Navy, 129

Sa Dec, 169, 175
Saigon, 54, 55, 57, 62, 71, 73, 74, 93, 101, 103, 108, 109, 114, 115, 118, 123, 140, 146, 152, 159, 171, 182, 188, 197, 200, 202, 208, 224, 228, 250

Cholon District, 110, 146
US Embassy, 70
'Y' Bridge, 87, 101, 103, 105, 108, 110, 115, 130
Scarborough, Captain Edmund B., 94
School of Air Navigation (SAN), 42, 45
Scott, Bruce, 284
Sheridan, Tony, 74
Sibthorpe, Arthur, 252
Silver City Airways, 28
Simon, Rex, 47
Singapore, 49, 54, 249
Smith, Flight Lieutenant Pete, 250, 252
Snoopy's Nose, 177, 179, 181, 197, 199
Stewart, Graham, 255
Sultan, Captain Anthony, 190
Swander, Bruce, 208

Taipei, 222, 223, 224, 225
Taiwan, 221
Tan An, 64, 65, 71, 74, 79, 80, 87, 89, 91, 93, 101, 102, 106, 108, 112, 114, 117, 119, 121, 126, 128, 130, 133, 137, 142, 143, 146, 150, 151, 153, 154, 155, 156, 157, 159, 160, 161, 176, 194, 196
Tan An Island, 150, 151, 152, 153, 154, 155, 156, 157, 158, 159
Tan Son Nhut Airbase, 54, 71, 90, 91, 104, 118, 121, 144, 152, 171, 202, 220, 225, 226, 239
Taylor, Lieutenant Charlie, 95
Territory Airlines Limited, 35
Tet Offensive, 55, 70, 71, 76, 81, 108
Thala Dan, 44, 46, 47
Thanh, General Ngoc Nha, 209
Thap Muoi Canal, 171, 182
Tibbett, Doctor, 182
Tiller, Sergeant Don, 46
Trans Australian Airlines, 38

Treloar, Bill, 28
Truc Giang, 191
Turner, Colonel Joe, 74, 110, 111
Turnnidge, Squadron Leader George, 34, 39, 41, 42, 60

Ubon, 49
United States Navy Aviation Museum, 121

Vam Co Dong River, 68, 88, 92, 121, 135, 140, 147
Vam Co Tay River, 68, 92, 102, 147, 150, 151, 152, 154, 189, 199
Van Deusen, Colonel Edward, 172
Van Deusen, Lieutenant Colonel Edwin, 172
Van Deusen, Lieutenant Colonel Frederick, 172
Van Dyck, Ron, 79
VC Island, 231
VC Panhandle, 170, 171, 176, 242
Vespico, Kathleen, 165
Viet Cong (VC), 3, 4, 5, 13, 58, 61, 62, 68, 70, 97, 102, 109, 128, 131, 132, 133, 134, 137, 140, 142, 144, 147, 148, 149, 150, 151, 153–158, 163, 165, 168, 171, 173, 176, 177, 183, 187, 192, 195, 197, 198, 208, 210–14, 231, 237, 242, 249
Vietnam, 2, 53, 72, 88, 161, 225, 250
  airforce (VNAF), 83–4
  army (ARVN), 5, 67, 129, 131, 132, 133, 169, 233
Vietnam Veterans' Museum, 88
Vietnamese placenames, 276

Vinh Long, 20, 167, 175, 235
Vostock, 46
Vung Tau, 161, 162, 163, 165, 182, 239, 245, 246, 248

Walker, 'Blackjack', 34
Walker, Major William 'Bill', 3, 64, 65, 66, 68, 108, 109, 117, 118, 130, 131, 135, 136, 139, 140, 143, 144, 146, 161, 163, 167, 169, 170, 171, 187, 265, 287
Washburn, Captain Don, 66, 107, 108, 109, 110, 126, 130, 143, 170, 178, 181, 196, 224, 226
Waterfield, Dick, 42
Watson, Flight Lieutenant Tex, 40, 41
Weaver, 62
Westmoreland, General, 160, 172
Wewak, 32
Wilkes Station, 46
Wilkins, Randolph R., 94
Williams, Squadron Leader 'Nobby', 250
Wills, Doc, 32
Wilson, Flight Lieutenant Roger, 54, 55, 57, 59, 62, 238, 248, 252
Wright, Graham, 29, 30

Xang Canal, 160, 186
Xom Chong, 90
Xom Tan Liem, 93, 94, 120, 124
Xom Tre, 236, 236, 237, 241, 244

Young, Flight Lieutenant Alec, 38

Zamboanga, 1